# Design and Analysis of
# Algorithms

# Design and Analysis of
# Algorithms

**Anam Faruqi**

Aligarh College of Engineering and
Technology

**CBS**

# CBS Publishers & Distributors Pvt Ltd

New Delhi • Bengaluru • Chennai • Kochi • Kolkata • Mumbai
Hyderabad • Nagpur • Patna • Pune • Vijayawada

Design and Analysis of
# Algorithms

**ISBN:** 978-93-85915-04-8

Copyright © Author and Publisher

**First Edition: 2016**

Published by Satish Kumar Jain and produced by Varun Jain for

**CBS Publishers & Distributors** Pvt Ltd

4819/XI Prahlad Street, 24 Ansari Road, Daryaganj, New Delhi 110 002, India.
Ph: 23289259, 23266861, 23266867   Website: www.cbspd.com
Fax: 011-23243014                  e-mail: delhi@cbspd.com; cbspubs@airtelmail.in.
*Corporate Office:* 204 FIE, Industrial Area, Patparganj, Delhi 110 092
Ph: 4934 4934       Fax: 4934 4935   e-mail: publishing@cbspd.com; publicity@cbspd.com

*Branches*

- **Bengaluru:** Seema House 2975, 17th Cross, K.R. Road,
  Banasankari 2nd Stage, Bengaluru 560 070, Karnataka
  Ph: +91-80-26771678/79        Fax: +91-80-26771680        e-mail: bangalore@cbspd.com
- **Chennai:** 7, Subbaraya Street, Shenoy Nagar, Chennai 600 030, Tamil Nadu
  Ph: +91-44-26680620, 26681266        Fax: +91-44-42032115        e-mail: chennai@cbspd.com
- **Kochi:** Ashana House, No. 39/1904, AM Thomas Road, Valanjambalam,
  Ernakulam 682 018, Kochi, Kerala
  Ph: +91-484-4059061-65        Fax: +91-484-4059065        e-mail: kochi@cbspd.com
- **Kolkata:** 6/B, Ground Floor, Rameswar Shaw Road, Kolkata-700 014, West Bengal
  Ph: +91-33-22891126, 22891127, 22891128        e-mail: kolkata@cbspd.com
- **Mumbai:** 83-C, Dr E Moses Road, Worli, Mumbai-400018, Maharashtra
  Ph: +91-22-24902340/41        Fax: +91-22-24902342        e-mail: mumbai@cbspd.com

*Representatives*

- **Hyderabad** 0-9885175004  • **Nagpur**  0-9021734563  • **Patna** 0-9334159340
- **Pune**  0-9623451994  • **Vijayawada** 0-9000660880

*Printed at:* Swastik Packagings, Patparganj Industrial Area, Delhi-110092

*to*
*my loving mother*
*Dr Roshan Ara and*
*my caring father*
*Prof Nafis Ahmad*
*Faruqi*

# Preface

The present book is a comprehensive approach to design and analysis of algorithms. On a broader spectrum, it focuses on structure, strategic design and analysis of algorithms followed by a succession of appropriate examples. Emphasis has been given on assumptions, elementary explanations and detailed sketches.

The entire book is divided into four parts. The first part discusses the basic definition of algorithm, different analyzing methods and asymptotic notations. The second part throws light on different data structures that are frequently used in algorithms. The third part gives an insight into various approaches that are used for designing the algorithms. The fourth part covers miscellaneous topics that are relevant to the study of algorithms.

Basic concepts like asymptotic notation, Fibonacci heaps, RB tree, data structure, knapsack problem, heap sort, radix sort, etc. have been explained in detail to help students develop new algorithms and suggest modifications in the existing ones.

A lucid style of writing is adopted throughout the book to make it innovative and student-friendly.

The intended readers are undergraduate and postgraduate students of various engineering and management courses.

**Anam Faruqi**

# Contents

# Chapter 1

# Algorithm

- An algorithm is a step by step list of instructions for solving a given problem.
- It takes the input and transforms it into the output.
- The efficiency of an algorithm can be determined by analyzing the amount of resources used by it. These resources are memory, computational time, transmission bandwidth etc.

  For maximum efficiency, the usage of the resources should be minimum.
- *Time complexity*:

  It is the amount of time required to execute an algorithm.
- *Space complexity*:

  It is the amount of memory required to execute an algorithm.

## 1.1 ASYMPTOTIC NOTATION

- Asymptotic notation is applied to functions that characterize the different aspects of algorithms like the running time of algorithms, the amount of space they use etc.

  Following are some of the asymptotic notations:

  i. Θ-*notation* (theta notation):

  For a given function $g(n)$, we define $\Theta(g(n))$ as the set of functions

  $\Theta(g(n)) = \{f(n)$: there exist positive constants $c_1, c_2$ and $n_0$, such that

  $$0 \leq c_1 g(n) \leq f(n) \leq c_2 g(n), \text{ for all } n \geq n_0\}$$

  A function $f(n)$ belongs to the set $\Theta(g(n))$, if there exist positive constants $c_1$ and $c_2$ such that it can be sandwiched between $c_1 g(n)$ and $c_2 g(n)$, for sufficiently large $n$.

1

ii. *O-notation* (big-oh notation):

For a given function $g(n)$, we define $O(g(n))$ as the set of functions

$O(g(n)) = \{f(n)$: there exist positive constants $c$ and $n_0$, such that

$$0 \le f(n) \le cg(n), \text{ for all } n \ge n_0\}$$

For all values $n$ at and to the right of $n_0$ the value of the function $f(n)$ is on or below $cg(n)$.

*O*-notation is used to give an upper bound on a function, to within a constant factor.

iii. *Ω-notation* (big-omega notation):

For a given function $g(n)$, we define $\Omega(g(n))$ as the set of functions

$\Omega(g(n)) = \{f(n)$: there exist positive constants $c$ and $n_0$ such that

$$0 \le cg(n) \le f(n), \text{ for all } n \ge n_0\}$$

For all values of $n$ at or to the right of $n_0$, the value of $f(n)$ is on or above $cg(n)$.

*Ω*-notation is used to give a lower bound on a function.

iv. *o-notation* (little-oh notation):

We define $o(g(n))$ as the set

$o(g(n)) = \{f(n)$: for any positive constant $c > 0$: there exists a constant $n_0 > 0$ such that

$$0 \le f(n) < cg(n), \text{ for all } n \ge n_0$$

In *o*-notation, the function $f(n)$ becomes insignificant relative to $g(n)$ as $n$ approaches infinity.

$$\lim_{n \to \infty} \frac{f(n)}{g(n)} = 0$$

*o*-notation is used to denote an upper bound which is not asymptotically tight.

v. *ω-notation* (little-omega notation):

We define $\omega(g(n))$ as the set

$\omega(g(n)) = \{f(n)$ : for any positive constant $c > 0$: there exists a constant $n_0 > 0$, such that

$$0 \le cg(n) < f(n) \text{ for all } n \ge n_0\}$$

In *ω*-notation, the function $f(n)$ becomes arbitrarily large relative to $g(n)$ as $n$ approaches infinity.

$$\lim_{n \to \infty} \frac{f(n)}{g(n)} = \infty \quad \text{(if limit exists).}$$

ω-notation is used to denote a lower bound which is not asymptotically tight.

*Note:*

- The time complexity of an algorithm is commonly expressed using 'big-*O* notation'.
- Usually the time complexity is estimated by counting the number of elementary operations performed by the algorithm. An elementary operation takes a fixed amount of time to perform. Thus, the amount of time taken and the number of elementary operations performed by the algorithms differ by at most a constant factor.
- The space complexity of an algorithm is also commonly expressed using big-*O* notation. It defines an upper bound on the data that the program can handle.
- Ω notations which give a lower bound on a function are helpful to determine whether it is possible to design a faster algorithm. The lower bound describes a certain number of steps that every algorithm has to execute at least in order to solve the problem.

## 1.2 WORST CASE, AVERAGE CASE AND BEST CASE ANALYSIS

### 1.2.1 Worst Case Analysis

- The worst case analysis of an algorithm gives an upper bound on the resources required by the algorithm.
- The worst case running time of an algorithm determines the maximum amount of time taken by the algorithm on any input.
- The worst case running time guarantees that the algorithm will never take any longer.

### 1.2.2 Average Case Analysis

- The average case analysis of an algorithm determines the amount of computational resource used by an algorithm, averaged over all possible inputs.
- The average case running time of an algorithm is an estimate of the running time for an average input.
- The computation of average running time requires the knowledge of all possible input sequences, the probability distribution of occurrence of these sequences and the running time for the individual sequences.

### 1.2.3 Best Case Analysis

- The best case analysis of an algorithm gives a lower bound on the resources required by the algorithm.
- The best case running time of an algorithm determines the minimum amount of time taken by the algorithm on any input.
- The best case running time ensures that the algorithm runs the fastest.

  In general worst case, average case and best case analyses are done to determine the running time of algorithms but they can be useful for determining the requirement of memory or other resources as well.

## 1.3 AMORTIZED ANALYSIS

- In an amortized analysis of running time of an algorithm, the time required to perform a sequence of data structure operations is averaged over all the operations performed.
- Amortized analysis can be used to show that the average cost of an operation is small, if one averages over a sequence of operations, even though a single operation within the sequence might be expensive.
- Amortized analysis differs from average case analysis in the sense that amortized analysis does not involve probability. Amortized analysis guarantees the average performance of each operation in the worst case.

### 1.3.1 Aggregate Analysis

- In an aggregate analysis, for all $n$, a sequence of $n$ operations takes worst case time $T(n)$ in total.
- Therefore, in the worst case, the average cost or amortized cost, per operation is $T(n)/n$. Here the amortized cost applies to each operation, even when there are several types of operations in the sequence.

### 1.3.2 The Potential Method

- The potential method represents the prepaid work as 'potential' which can be released to pay for future operations.
- This potential is associated with the data structure as a whole rather than with specific objects within the data structure.
- Let $n$ operations are performed, starting with an initial data structure $D_0$.

For each $i = 1, 2, \ldots, n$, let $c_i$ be the actual cost of the $i$th operation and $D_i$ be the data structure that results after applying the $i$th operation to data structure $D_{i-1}$.

A potential function $\varphi$ maps each data structure $D_i$ to a real number $\varphi(D_i)$, which is the potential associated with data structure $D_i$.

The amortized cost $(\hat{c}_i)$ of the $i$th operation with respect to potential function $\varphi$ is defined by $(\hat{c}_i) = c_i + \varphi(D_i) - \varphi(D_i - 1)$

Therefore, the amortized cost of each operation is its actual cost plus the change in potential due to the operation.

The total amortized cost of the $n$ operations is

$$\sum_{i=1}^{n} \hat{c}_i = \sum_{i=1}^{n} \left( c_i + \varphi(D_i) - \varphi(D_{i-1}) \right)$$

# Chapter 2

# Data Structure

- In computer science, a data structure is a particular way of organizing data in a computer so that it can be used efficiently.
- Efficiency is usually measured by two factors:
  i. Time
  ii. Space
- If a particular application is heavily dependent on manipulating high-level data structures, the speed at which those manipulations can be performed will be the major determinant of the speed of the entire application.
- On the other hand, if a particular application uses a large number of high level data structures, the amount of space required will be large.
- Thus, an implementation which is fast, uses more storage than one which is slow.
- So, usually there is a trade-off between time and space.
  Different kinds of data structures are suited to different kinds of applications.
- Examples of data structures are arrays, queues, linked lists, stacks, binary trees, red-black trees, B-trees, Fibonacci heaps, graphs etc.

| Data Structure | Advantages | Disadvantages |
| --- | --- | --- |
| 1. Array | Quick inserts<br>Fast access, if index is known | Slow search<br>Slow deletes<br>Fixed size |
| 2. Queue | First-in-first-out (FIFO) access | Slow access to other items |

| Data Structure | Advantages | Disadvantages |
|---|---|---|
| 3. Linked lists | Quick inserts<br>Quick deletes | Slow search |
| 4. RB trees | Quick search<br>Quick inserts<br>Quick deletes<br>(Tree always remains balanced) | Complex to implement |
| 5. Heap | Quick inserts<br>Quick deletes<br>Access to largest item | Slow access to other items |
| 6. Graph | Best models real world situations | Some algorithms are slow and complex |

Algorithms manipulate the data contained in data structures for different purposes like:

   i.  Searching for a particular data item
   ii. Sorting the data
   iii. Iterating through all the items in a data structure

## 2.1 LINEAR ARRAYS

- A linear array is a list of a finite number $n$ of homogeneous data elements.
- The elements of an array are referenced respectively by an index set consisting of $n$ consecutive numbers.
- The elements of an array are stored respectively in successive memory locations.
- The number $n$ of elements is called the length or size of an array.

  Following figure shows an array of 5 elements:

| 1 | 2 | 3 | 4 | 5 |
|---|---|---|---|---|
| 5 | 8 | 12 | 15 | 20 |

## 2.2 QUEUES

- A queue is an ordered collection of items.
- In a queue the insertions can take place at one end called the 'rear' and the deletions can take place at the other end called the 'front'.
- Queues are known as 'first-in-first-out' (FIFO) lists because the order in which the elements enter a queue is the order in which they leave.

## Algorithm

Enqueue (Q, x)
Line 1       Q [tail [Q]] ← x
Line 2       if   tail [Q] = = length [Q]
Line 3                 tail [Q] ←  1
Line 4       else  tail [Q] ← tail [Q] + 1

## Analysis

- Enqueue procedure takes $O(1)$ time.

## Explanation

- The inputs to the Enqueue procedure are a queue $Q$ and an element $x$ which is to be inserted into the queue $Q$.
- Line 1 indicates that the element $x$ is inserted at the tail of the queue. It means $x$ is inserted at the rear end of queue.
- Line 2 checks the *if* condition. If this condition is true then the execution of Line 3 takes place, else the control goes to Line 4. The *if* condition is true when the attribute tail [Q] is equal to the length of Q. It means tail [Q] points to the last location of queue $Q$.
- Line 3 indicates that the attribute tail [Q] is made to point to the first location of $Q$.
- Line 4 indicates that the attribute tail [Q] is made to point to the next location.

## Example

Illustrate the result of each operation on the following queue $Q$:
i. Enqueue  $(Q, 10)$
ii. Enqueue $(Q, 5)$

|   | 1 | 2 | 3 | 4 | 5 | 6 | 7 |
|---|---|---|---|---|---|---|---|
| Q |   |   | 12 | 8 | 4 | 9 |   |

## Solution

length [Q] = 7

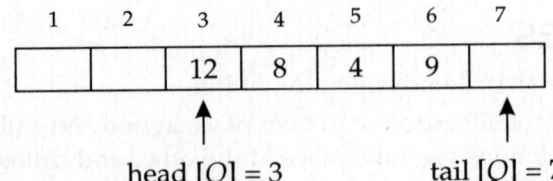

head [Q] = 3                        tail [Q] = 7

i. Enqueue (Q, 10)

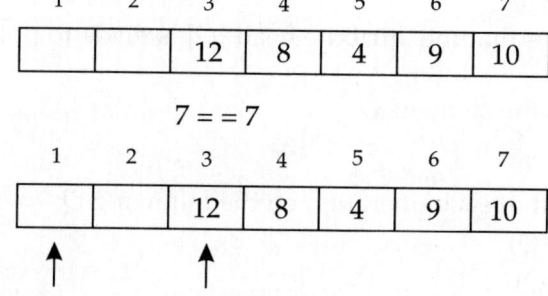

7 = = 7

tail [Q] = 1      head [Q] = 3

ii. Enqueue (Q, 5)

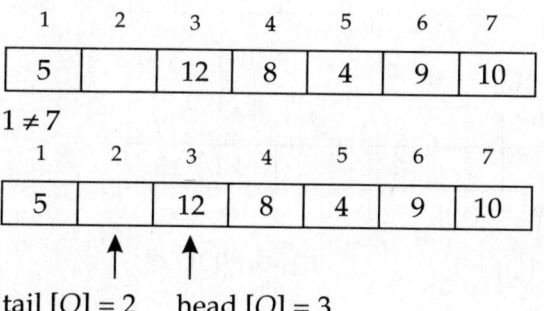

1 ≠ 7

tail [Q] = 2    head [Q] = 3

## Algorithm

```
        Dequeue (Q)
Line 1          x ← Q [head [Q]]
Line 2          if  head [Q] = = length [Q]
Line 3                  head [Q] ← 1
Line 4          else   head [Q] ← head [Q] + 1
Line 5          return  x
```

## Analysis

- The Dequeue procedure takes $O(1)$ time.

## Explanation

- The input to the Dequeue procedure is a queue $Q$. This procedure is used to delete elements from the queue.
- Line 1 indicates that $x$ is the element at location $Q[head [Q]]$.
- Line 2 checks the *if* condition. If this condition is true then the execution of Line 3 takes place else the control goes to Line 4.
- The *if* condition is true when the attribute head [Q] is equal to the length of Q. It means head [Q] points to the last location of Q.

- Line 3 indicates that the attribute head [Q] is made to point to the first location of Q.
- Line 4 indicates that the attribute head [Q] is made to point to the next location.
- Line 5 returns the element x.

## Example

Illustrate the result of each operation on the following Q:

    i.  Dequeue (Q)

   ii.  Dequeue (Q)

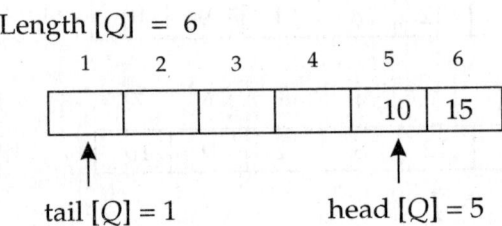

## Solution

$$\text{Length } [Q] = 6$$

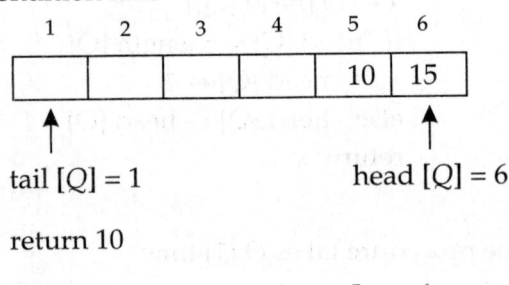

  i.  Dequeue (Q)

$$x = 10$$
$$5 \neq 6$$

So, *if* condition fails

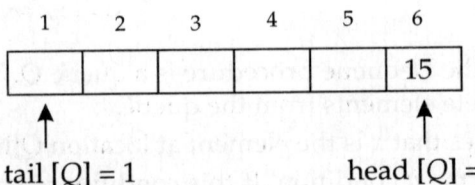

return 10

  ii.  Dequeue (Q)

$$x = 15$$
$$6 = = 6$$

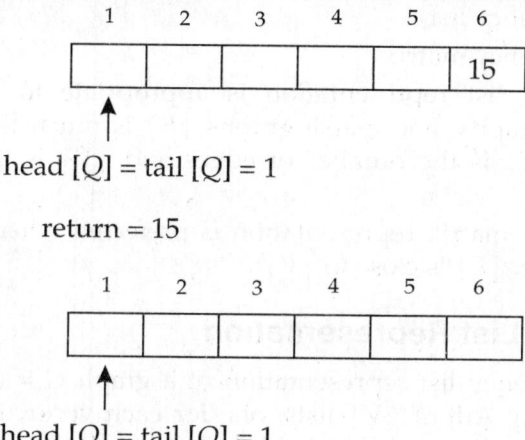

head [Q] = tail [Q] = 1

return = 15

head [Q] = tail [Q] = 1

## 2.3 LINKED LISTS

- A linked list is a linear collection of nodes.
- Each node is divided into two parts. The first part contains the information of the element and the second part contains the address of the next node in the list.

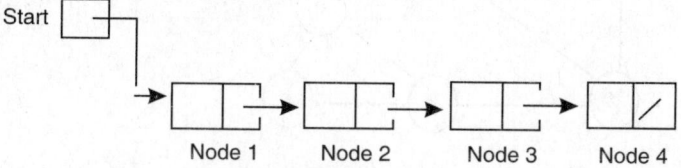

In the above figure, each node has two parts. The left part represents the information part and the right part represents the next pointer field of the node. An arrow is drawn from the right part of the node to the next node in the list. The right part of the last node contains a null pointer. It indicates the end of the list. The linked list contains a list pointer variable 'Start' which contains the address of the first node in the list.

## 2.4 GRAPHS

A graph G consists of a set V of vertices and a set E of edges.

- If the pairs of nodes that make up edges are ordered pairs, then the graph is said to be a directed graph.
- A graph data structure may also associate to each edge, some edge value such as weight, cost, capacity, length etc.
- A graph can be represented in two ways:

   i.    Adjacency list

  ii.    Adjacency matrix

- Adjacency list representation is appropriate to represent the 'sparse' graphs. For sparse graphs $|E|$ is much less than $|V|^2$, where, $|E|$ is the number of edges and $|V|$ is the number of vertices.

- Adjacency matrix representation is preferred when the graph is 'dense', i.e. $|E|$ is close to $|V|^2$.

## Adjacency List Representation

- The adjacency list representation of a graph $G = (V, E)$ consists of an array Adj of $|V|$ lists, one for each vertex in $V$. For each $u \in V$, the adjacency list Adj $[u]$ contains all the vertices $v$ such that there is an edge $(u, v) \in E$. That is, Adj $[u]$ consists of all the vertices adjacent to $u$ in $G$.

- The amount of memory required by adjacency list representation for both directed and undirected graphs is $\Theta (|V| + |E|)$.

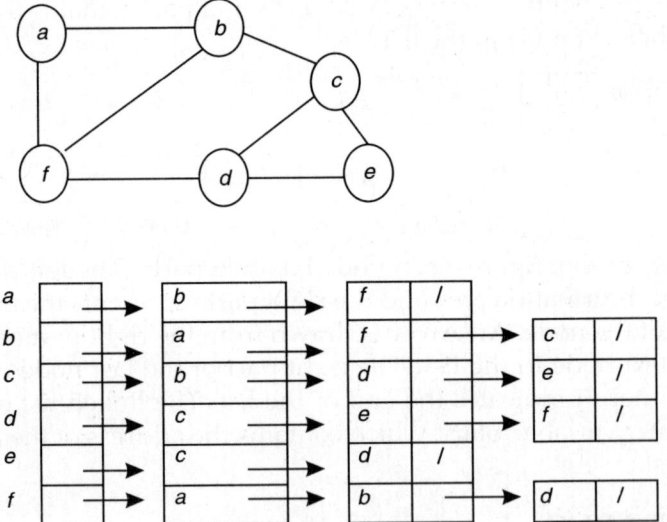

    The above figure shows an undirected graph and its adjacency list representation.

## 2.5 RED BLACK (RB) TREES

- A red-black tree is a binary tree with one extra bit of storage per node. This extra bit stores the color of the node which can be either red or black. In case if a node is 'doubly black', its, color attribute will be black. If a node is 'red and black', its color attribute will be red.

- Each node of the tree contains the attributes color, key, left, right and $P$.
- left, right and $P$ attributes of a node, point to its left child, its right child and its parent respectively.
- The keys in a RB tree are stored in such a way that they satisfy the following binary search tree property.

    Let $x$ is a node. If $y$ is a node in the left subtree of $x$, then key $[y] \le$ key $[x]$. If $y$ is a node in the right subtree of $x$, then key $[y] \ge$ key $[x]$.

- For a red-black tree $T$, the sentinel nil $[T]$ is an object with the same attributes as an ordinary node in the tree. Its color attribute is black, and its other attributes, i.e. $P$, left, right, key can take on arbitrary values. This sentinel nil $[T]$ is used to represent all the nils which are either all leaves or the root's parent.
- A red-black tree satisfies the following red-black properties:
  1. Every node is either red or black.
  2. The root is black.
  3. Every leaf is black.
  4. If a node is red, then both its children are black.
  5. For each node, all simple paths from the node to descendent leaves contain the same number of black nodes.

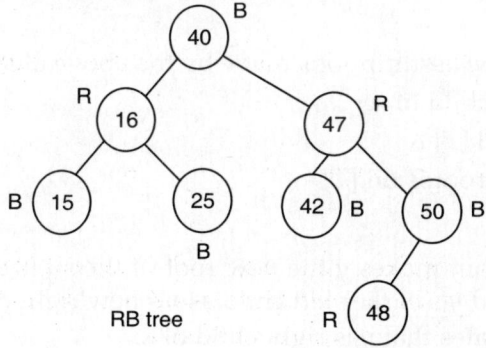

RB tree

- *Black-height* of the node $x$ is the number of black nodes on any simple path from (but not including) a node $x$ down to a leaf.
- The black-height of a red-black tree is the black-height of its root.
- A red-black tree with $n$ internal nodes has height at most $2[\lg (n+1)]$.

## Rotation

- Rotation is a local operation in a search tree.
- Rotation changes the pointer structure which is done to preserve the binary search tree properties.

Two types of rotations are discussed here:

(*i*) Left rotation

(*ii*) Right rotation

## Algorithm

Left-Rotate (T, x)

Line 1       $y \leftarrow$ right [x]

Line 2       right [x] $\leftarrow$ left [y]

Line 3       if   left [y] $\neq$ nil [T]

Line 4            P [left [y]] $\leftarrow$ x

Line 5       P [y] $\leftarrow$ P [x]

Line 6       if   P [x] = = nil [T]

Line 7            root [T] $\leftarrow$ y

Line 8       else if   x = = left [P [x]]

Line 9            left [P[x]] $\leftarrow$ y

Line 10      else     right [P[x]] $\leftarrow$ y

Line 11      left [y] $\leftarrow$ x

Line 12      P [x] $\leftarrow$ y

## Assumptions

Following are the assumptions made by the above algorithm:

1. $y$ is right child of $x$
2. $y$ is not nil [$T$]
3. Parent of root is nil [$T$]

## Explanation

- This algorithm makes $y$ the new root of the subtree with $x$ as $y$'s left child and $y$'s earlier left child as $x$'s new right child.
- Line 1 indicates that $y$ is right child of $x$.
- Line 2 makes $y$'s earlier left child as $x$'s new right child.
- Line 3 checks the *if* condition. This condition is true if left [$y$] is not nil [$T$]. If the condition is true, then the execution of Line 4 takes place. If this condition is false then the procedure does not execute Line 4 and the control goes to Line 5.
- Line 4 makes $x$ as the parent of the $y$'s left subtree.
- Line 5 links the $x$'s parent to $y$.

- Line 6 checks the *if* condition. This condition is true if the parent of $x$ is nil [$T$]. If this condition is true then the execution of Line 7 takes place else the control goes to Line 8.
- Line 7 makes $y$ as the root [$T$].
- Line 8 checks the *else if* condition. This condition is true if $x$ and left [$P$ [$x$]] are same. If this condition is true then the execution of Line 9 takes place else the execution of Line 10 takes place.
- Line 9 makes $y$ as the left subtree of $x$'s parent.
- Line 10 makes $y$ as the right subtree of $x$'s parent.
- Line 11 puts $x$ on $y$'s left.
- Line 12 makes $y$ as the parent of $x$.

**Analysis**

- This procedure takes $O(1)$ time.

**Example**

Apply Left-Rotate $(T, x)$ procedure on the following binary tree:

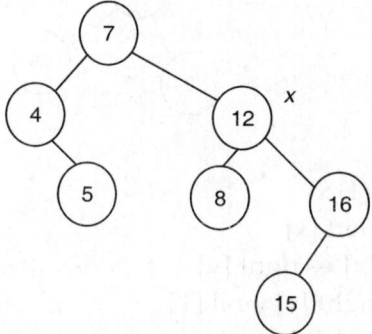

**Solution**

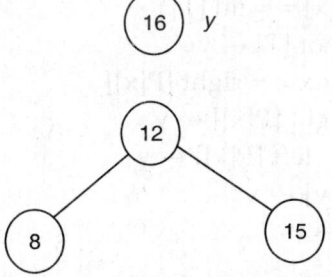

In this case, the *if* condition of Line 6 and the *elseif* condition of Line 8 both are false.

After the execution of Line 10

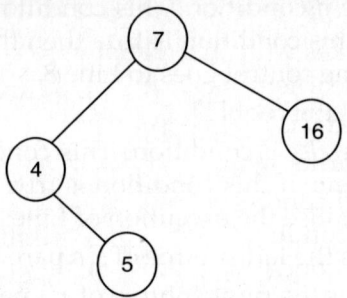

After the execution of Line 11 and Line 12

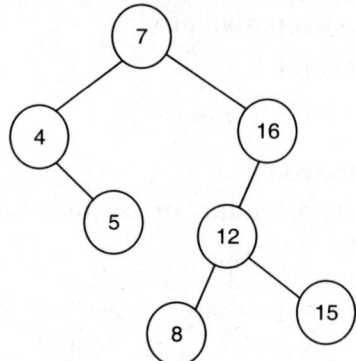

## Algorithm

Right-Rotate (T, x)
Line 1        y ← left [x]
Line 2        left [x] ← right [y]
Line 3        if   right [y] ≠ nil [T]
Line 4              P [right [y]] ← x
Line 5        P [y] ← P [x]
Line 6        if   P [x] = = nil [T]
Line 7              root [T] ← y
Line 8        elseif   x = = right [P[x]]
Line 9              right [P[x]] ← y
Line 10      else      left [P[x]] ← y
Line 11      right [y] ← x
Line 12      P [x] ← y

## Assumptions

Following are the assumptions made from the above algorithm:
  i. y is left child of x
 ii. y is not nil [T]
iii. Parent of root is nil [T]

## Explanation

- This algorithm makes $y$ the new root of the subtree with $x$ as $y$'s right child and $y$'s earlier right child as $x$'s new left child.
- Line 1 indicates that $y$ is left child of $x$.
- Line 2 makes $y$'s earlier right child as $x$'s new left child.
- Line 3 checks the *if* condition. This condition is true if right $[y]$ is not nil $[T]$. If the condition is true, then the execution of Line 4 takes place. If this condition is false then the procedure does not execute Line 4 and the control goes to Line 5.
- Line 4 makes $x$ as the parent of the $y$'s right subtree.
- Line 5 links the $x$'s parent to $y$.
- Line 6 checks the *if* condition. This condition is true if the parent of $x$ is nil $[T]$. If this condition is true then the execution of Line 7 takes place else the control goes to Line 8.
- Line 7 makes $y$ as the root $[T]$
- Line 8 checks the *elseif* condition. This condition is true if $x$ and right $[P[x]]$ are same. If this condition is true then the execution of Line 9 takes place else the execution of Line 10 takes place.
- Line 9 makes $y$ as the right subtree of $x$'s parent.
- Line 10 makes $y$ as the left subtree of $x$'s parent.
- Line 11 puts $x$ on $y$'s right.
- Line 12 makes $y$ as the parent of $x$.

## Analysis

- The above procedure takes $O(1)$ time.

## Example

Apply Right-Rotate $(T, x)$ procedure on the following binary tree.

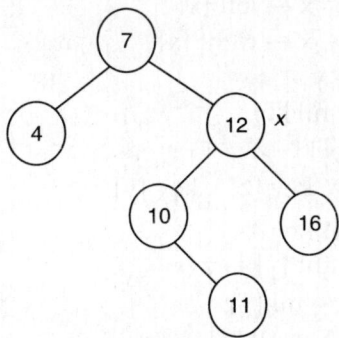

## Solution

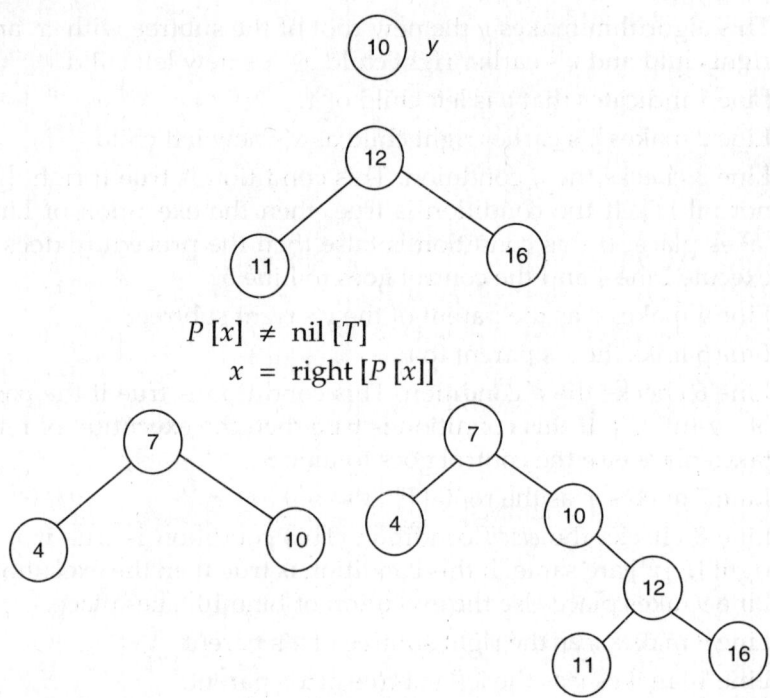

$$P[x] \neq nil[T]$$
$$x = right[P[x]]$$

## Algorithm

    RB-Insert (T, z)
    Line 1      y ← nil [T]
    Line 2      x ← root [T]
    Line 3      while  x ≠ nil [T]
    Line 4          y ← x
    Line 5          if  key [z] < key [x]
    Line 6              x ← left [x]
    Line 7          else  x ← right [x]
    Line 8      P [z] ← y
    Line 9      if y = = nil [T]
    Line 10        root [T] ← z
    Line 11     else if   key [z] < key [y]
    Line 12         left [y] ← z
    Line 13     else  right [y] ← z
    Line 14     left [z] ← nil [T]
    Line 15     right [z] ← nil [T]
    Line 16     color [z] ← Red
    Line 17     RB-Insert-Fixup (T, z)

## Explanation

The above algorithm inserts node $z$, whose key is assumed to have already been filled in, into the red-black tree $T$.

- Line 1 sets pointer $y$ to nil [$T$].
- Line 2 indicates that pointer $x$ is made to point to the root of tree $T$.
- Line 3 indicates the beginning of while loop that ends with Line 7. This loop terminates when $x$ becomes nil [$T$]. This while loop causes the two pointers $x$ and $y$ to move down the tree, going left or right depending on the comparison of key [$z$] with key [$x$].
- Line 4 indicates that the pointer $y$ is made to point to that node which is pointed by pointer $x$.
- Line 5 checks the *if* condition. This condition is true when key of $x$ is greater than key of $z$. If this condition is true then the execution of Line 6 takes place else the execution of Line 7 takes place.
- Line 6 indicates that the pointer to node $x$ is now made to point to the left child of node $x$.
- Line 7 indicates that the pointer to node $x$ is made to point to the right child of node $x$.
- Line 8 indicates that the parent of $z$ is that node which is pointed by pointer $y$.
- Line 9 checks the *if* condition. This condition is true when $y$ is pointing to nil [$T$]. If this condition is true then the execution of Line 10 takes place otherwise the procedure checks the *elseif* condition of Line 11.
- Line 10 indicates that node $z$ becomes the root of tree $T$.
- Line 11 indicates that the *elseif* condition is true when key of $y$ is greater than the key of $z$. If this condition is false then the execution of Line 13 takes place.
- Line 12 indicates that node $z$ becomes the left child of node $y$.
- Line 13 indicates that node $z$ becomes the right child of node $y$.
- Line 14 indicates that left child of $z$ is set to nil [$T$].
- Line 15 indicates that right child of $z$ is set to nil [$T$].
- Line 16 indicates that the color of node $z$ is set to red.
- Line 17 calls the RB-Insert-Fixup ($T$, $z$) procedure.

## Algorithm

RB-Insert-Fixup (T, z)

Line 1     while   color [P [z]] = = Red
Line 2          if   P [z] = = left [P [P [z]]]
Line 3               y ← right [P [P [z]]]
Line 4               if   color [y] = = Red

Case 1 {
Line 5                    color [P [z]] ← Black
Line 6                    color [y] ← Black
Line 7                    color [P [P [z]]] ← Red
Line 8                    z ← P [P [z]]
}

Line 9               else if   z = = right [P [z]]

Case 2 {
Line 10                       z ← P [z]
Line 11                       Left-Rotate (T, z)
}

Case 3 {
Line 12                    color [P[z]] ← Black
Line 13                    color [P [P [z]]] ← Red
Line 14                    Right-Rotate (T, P [P [z]])
}

Line 15          else   y ← left [P [P [z]]]
Line 16               if   color [y] = = Red

Case 4 {
Line 17                    color [P [z]] ← Black
Line 18                    color [y] ←   Black
Line 19                    color [P [P [z]]] ← Red
Line 20                    z ← P [P [z]]
}

Line 21               else if   z = = left [P [z]]

Case 5 {
Line 22                       z ← P [z]
Line 23                       Right-Rotate (T, z)
}

Case 6 {
Line 24                    color [P [z]] ← Black
Line 25                    color [P [P [z]]] ← Red
Line 26                    Left-Rotate (T, P [P [z]])
}

Line 27     color [root [T]] ← Black

## Explanation

- Line 1 indicates the beginning of while loop that ends with Line 26. This while loop is applicable until the color of the parent of node $z$ remains red.
- Line 2 checks the *if* condition. This condition is true when parent of $z$ is the left child of the parent of parent of $z$. If this condition is false then the control goes to Line 15.

- Line 3 makes pointer $y$ to point to the right child of the parent of parent of $z$. It means that $y$ points to the right child of the grandparent of $z$, or in other words $y$ points to $z$'s uncle.
- Line 4 checks the *if* condition. This condition is true if the color of node $y$ is red. If this condition is true then the execution of Lines 5, 6, 7 and 8 takes place. If this condition is false then the procedure checks the *elseif* condition of Line 9.
- Line 5 paints the parent of $z$ black.
- Line 6 paints node $y$ black.
- Line 7 paints grandparent of $z$ red.
- Line 8 makes the earlier grandparent of $z$ as the new $z$. So the pointer $z$ moves up two levels in the tree.
- Line 9 checks the *elseif* condition. This condition is true if $z$ is the right child of the parent of $z$. If this condition is false then the control goes to Line 12.
- Line 10 makes the earlier parent of $z$ as the new $z$. So the pointer $z$ moves one level up in the tree.
- Line 11 calls the procedure Left-Rotate $(T, z)$.
- Line 12 paints the parent of $z$ black.
- Line 13 paints the grandparent of $z$ red.
- Line 14 calls the procedure Right-Rotate $(T, P [P[z]])$.
- Line 15 makes pointer $y$ to point to the left child of the parent of parent of $z$. It means that $y$ points to the left child of the grandparent of $z$ or in other words $y$ points to $z$'s uncle.
- Line 16 checks the *if* condition. This condition is true if the color of node $y$ is red. If this condition is true then the execution of Lines 17, 18, 19 and 20 takes place. If this condition is false then the procedure checks the *elseif* condition of Line 21.
- Line 17 paints the parent of $z$ black.
- Line 18 paints the node $y$ ($z$'s uncle) black.
- Line 19 paints grandparent of $z$ red.
- Line 20 makes the earlier grandparent of $z$ as the new $z$. So the pointer $z$ moves up two levels in the tree.
- Line 21 checks the *elseif* condition. This condition is true if $z$ is the left child of the parent of $z$. If this condition is false then the control goes to Line 24.
- Line 22 makes the earlier parent of $z$ as the new $z$. So the pointer $z$ moves one level up in the tree.
- Line 23 calls the procedure Right-Rotate $(T, z)$.
- Line 24 paints the parent of $z$ black.

- Line 25 paints the grandparent of $z$ red.
- Line 26 calls the procedure Left-Rotate $(T, P [P[z]])$.
- Line 27 paints the root of the tree $T$ black.

❑ During the execution of the procedure RB-Insert $(T, z)$ red black properties 2 and 4 may be violated. To restore these red-black properties, RB-Insert-Fixup $(T, z)$ procedure is called.

Property 2 says that the root of RB tree is black.

Property 4 says that if a node is red, then both its children are black.

The violation of these two properties occurs due to $z$ being colored red during the execution of procedure RB-Insert $(T, z)$.

Property 2 is violated if $z$ is the root.

Property 4 is violated if $z$'s parent is red.

❑ For better understanding of the pseudo code of RB-Insert-Fixup $(T, z)$, it is divided into six cases. Cases 4, 5, 6 are same as cases 1, 2, 3 respectively with 'left' and 'right' exchanged.

Once we enter into case 2, then we have to enter into case 3. So, we can enter into case 3 either directly or through case 2.

Similarly, once we enter into case 5, we have to enter into case 6. So we can enter into case 6 either directly or through case 5.

❑ If parent of $z$ is the left child of the grandparent of $z$, then case 1, case 2 and case 3 are taken into account.

❑ If parent of $z$ is the right child of the grandparent of $z$, then case 4, case 5 and case 6 are taken into account.

❑ Case 1 occurs when $z$'s uncle $y$ is red in color. Line 5 to Line 8 constitutes case 1.

❑ Case 2 occurs when $z$'s uncle $y$ is black in color and $z$ is a right child of its parent. Line 10 to Line 11 constitutes case 2.

❑ Case 3 occurs when $z$'s uncle $y$ is black in color and $z$ is a left child of its parent. Line 12 to Line 14 constitutes case 3.

❑ Case 4 occurs when $z$'s uncle $y$ is red in color. Line 17 to Line 20 constitutes case 4.

❑ Case 5 occurs when $z$'s uncle $y$ is black in color and $z$ is a left child of its parent. Line 22 to Line 23 constitutes case 5.

❑ Case 6 occurs when $z$'s uncle $y$ is black in color and $z$ is a right child of its parent. Line 24 to Line 26 constitutes case 6.

**Analysis**

- Since the height of a RB tree of $n$ nodes is $O (\lg n)$, Lines 1 to 16 of RB-Insert take $O (\lg n)$ time.

- In RB-Insert-Fixup, the while loop repeats only if case 1 occurs, and then the pointer $z$ moves two levels up the tree. The total number of times the while loop can be executed is therefore $O(\lg n)$.
- Thus, RB-Insert takes a total of $O(\lg n)$ time. Moreover, it never performs more than two rotations, since the while loop terminates, if case 2 or case 3 is executed.

## Example

Show the red-black trees after successively inserting the keys 41, 38, 31, 12, 19, 48, 46 into an initially empty red-black tree.

## Solution

First consider the node with key value 41

RB-Insert $(T, z)$

$$y = \text{nil}[T]$$
$$x = \text{nil}[T]$$

So, while loop will not execute

$$P[z] = \text{nil}[T]$$
$$y = = \text{nil}[T]$$

$z$   (41)   R

RB-Insert-Fixup $(T, z)$

color $[P[z]] \neq \text{Red}$

So, while loop will not execute

Now consider the node with key value 38.

$$y = \text{nil}[T]$$

$x$ points to node 41

$y$ points to node with key 41

$$38 < 41$$

$$x = \text{nil}[T]$$

Now, while loop terminates.

$P[z]$ points to the node with key 41

$$38 < 41$$

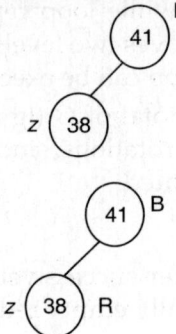

RB-Insert-Fixup $(T, z)$

    color $[P [z]] \neq$ Red

So, while loop will not execute

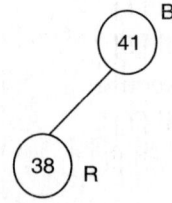

Consider the node with key value 31

$$y = \text{nil} [T]$$

$x$ points to node 41

$y$ points to  node 41

$$31 < 41$$

$x$ points to node  38

$$x \neq \text{nil} [T]$$

$y$ points to node  38

$$31 < 38$$

$$x = \text{nil} [T]$$

while loop terminates.

$P [z]$ is node 38

$$y \neq \text{nil} [T]$$

$$31 < 38$$

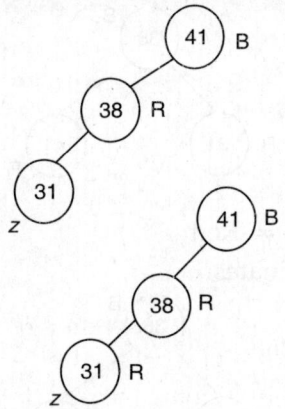

RB-Insert-Fixup $(T, z)$

    color $[P [z]] = =$ Red
    $P [z] = =$ left $[P [P [z]]]$
    $y =$ right $[P [P[z]]]$

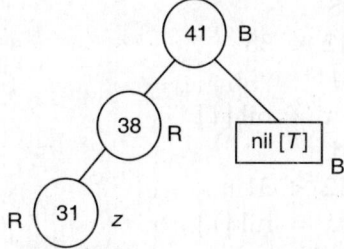

    color $[y] \neq$ Red
    $z \neq$ right $[P[z]]$

So, case 3 will be applied

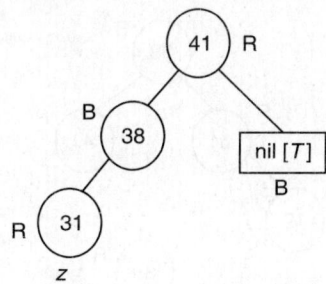

Right-Rotate $(T, P [P [z]])$

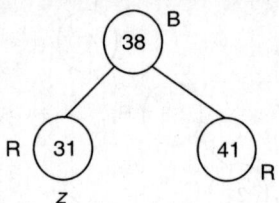

color $[P[z]] \neq$ Red

So, while loop terminates

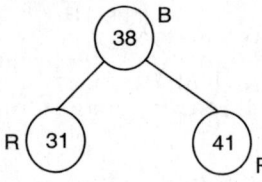

Next consider the node with key value 12

$$y = nil[T]$$

$x$ points to node 38

$y$ points to node 38

$$12 < 38$$

$x$ points to node 31

$$x \neq nil[T]$$

$y$ points to node 31

$$12 < 31$$

$$x = nil[T]$$

So, while loop terminates

$P[z]$ is node 31

$$y \neq nil[T]$$

$$12 < 31$$

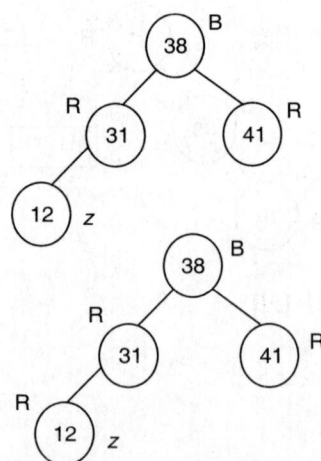

RB-Insert-Fixup $(T, z)$

$$\text{color } [P [z] ] = = \text{Red}$$
$$P [z] = = \text{left } [P [P [z]]]$$
$$y = \text{right } [P [P [z]]]$$

So, $y$ points to node 41

$$\text{color } [y] = = \text{Red}$$

Here, case 1 will be applied

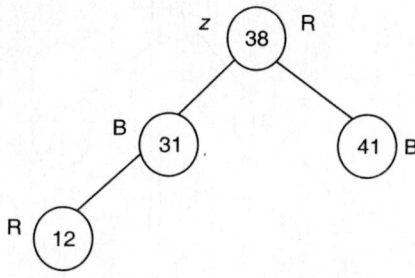

$$\text{color } [P [z]] \neq \text{Red}$$

So, while loop terminates

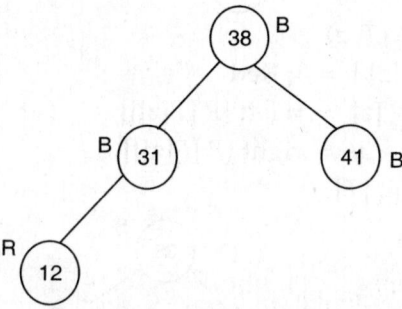

Now consider the node with key value 19

$$y = \text{nil } [T]$$

$x$ points to node 38

$$x \neq \text{nil } [T]$$

$y$ points to node 38

$$19 < 38$$

$x$ points to node 31

$$x \neq \text{nil } [T]$$

$y$ points to node 31

$$19 < 31$$

$x$ points to node 12

$$x \neq \text{nil } [T]$$

$y$ points to node 12

$$19 \nless 12$$

$x$ points to nil $[T]$

$$x = = \text{nil } [T]$$

So, while loop terminates

$P[z]$ is node 12

$$y \neq \text{nil } [T]$$
$$19 \nless 12$$

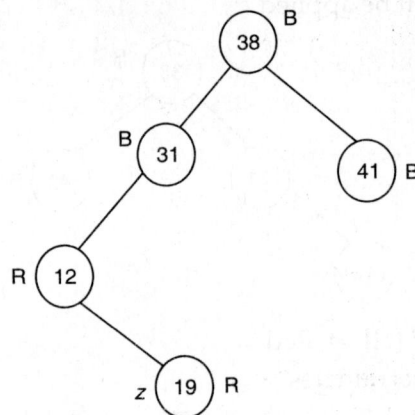

RB-Insert-Fixup $(T, z)$

$$\text{color } [P[z]] = = \text{Red}$$
$$P[z] = = \text{left } [P[P[z]]]$$
$$y = \text{right } [P[P[z]]]$$

So, $y$ points to nil $[T]$

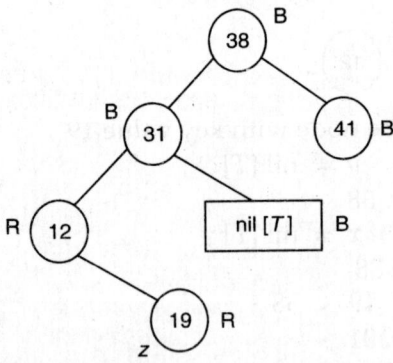

$$\text{color } [y] \neq \text{Red}$$
$$z = = \text{right } [P[z]]$$

So, case 2 will be applied

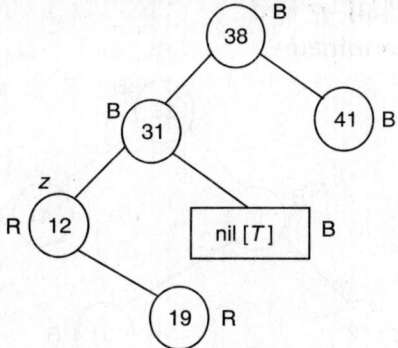

Left-Rotate $(T, z)$

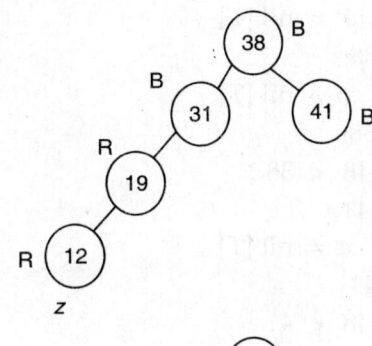

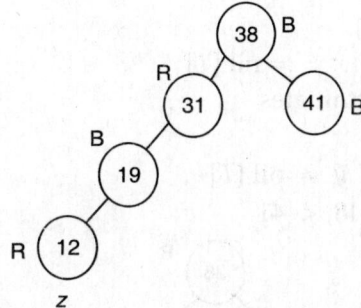

Right-Rotate $(T, P [P[z]])$

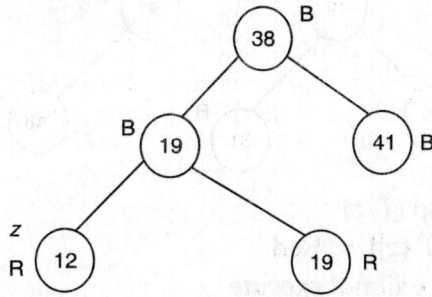

color $[P[z]] \neq$ Red

So, while loop terminates

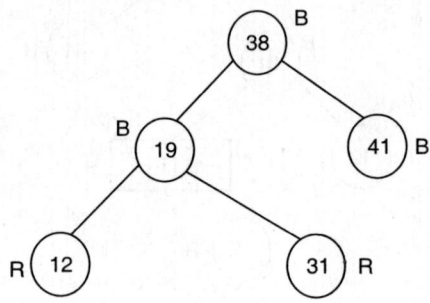

Now consider the node with key value 48

$$y = nil [T]$$

$x$ points to node 38

$$x \neq nil [T]$$

$y$ points to node 38

$$48 \nless 38$$

$x$ points to node 41

$$x \neq nil [T]$$

$y$ points to node 41

$$48 \nless 41$$

$x$ points to nil $[T]$

$$x = = nil [T]$$

So, while loop terminates

$P[z]$ is node 41

$$y \neq nil [T]$$
$$48 \nless 41$$

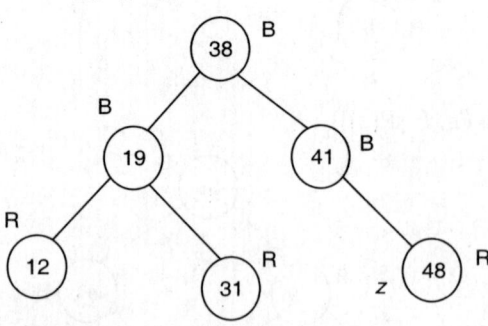

RB-Insert-Fixup $(T, z)$

color $[P[z]] \neq$ Red

So, while loop will not execute

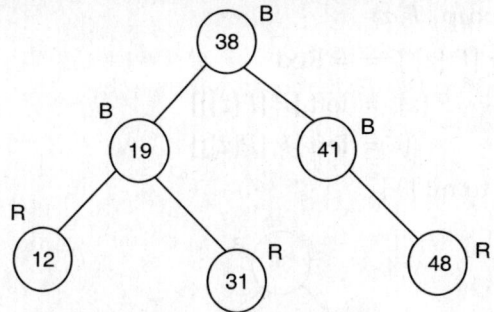

Now consider the node with key value 46

$$y = nil [T]$$

$x$ points to node 38

$$x \neq nil [T]$$

$y$ points to node 38

$$46 \not< 38$$

$x$ points to 41

$$x \neq nil [T]$$

$y$ points to 41

$$46 \not< 41$$

$x$ points to 48

$$x \neq nil [T]$$

$y$ points to 48

$$46 < 48$$

$x$ points to nil $[T]$

$$x = = nil [T]$$

So, while loop terminates

$P[z]$ is node 48

$$y \neq nil [T]$$
$$46 < 48$$

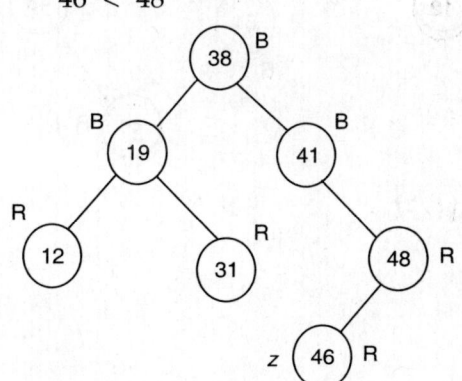

RB-Insert-Fixup $(T, z)$

      color $[P\,[z]\,]\ =\ =$ Red

           $P\,[z]\ \neq\ $ left $[P\,[P[z]]]$

             $y\ =\ $ left $[P\,[P[z]]]$

So, $y$ points to nil $[T]$

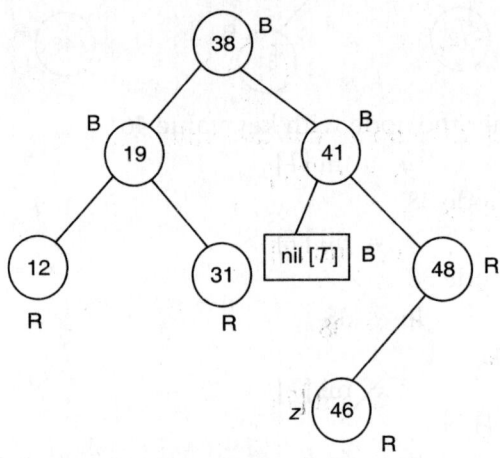

      color $[y]\ \neq\ $ Red

           $z\ =\ =$ left $[P\,[z]]$

So, case 5 will be applied

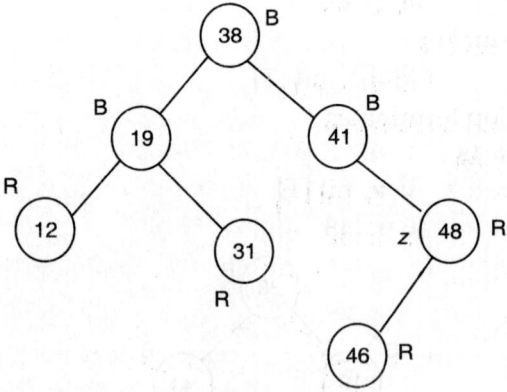

Right-Rotate $(T, z)$

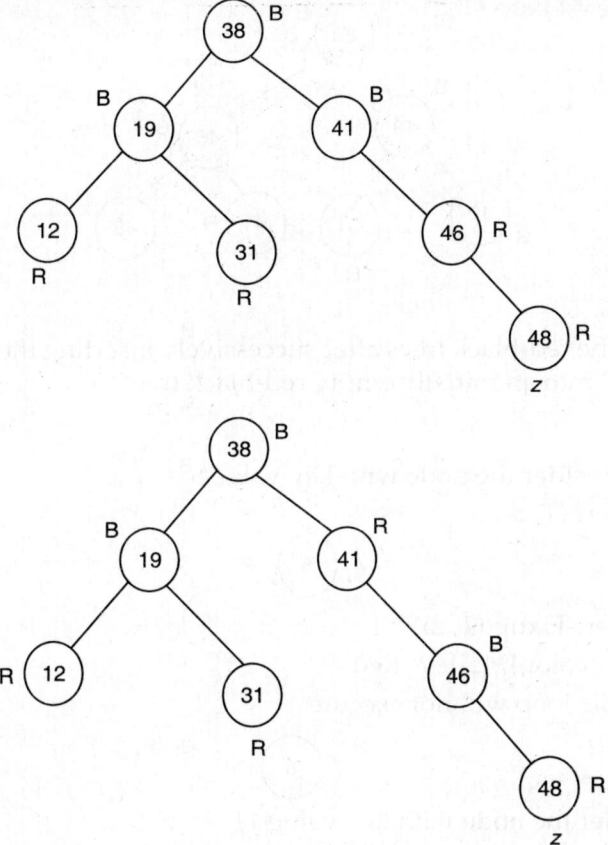

Left-Rotate (T, P [P[z]])

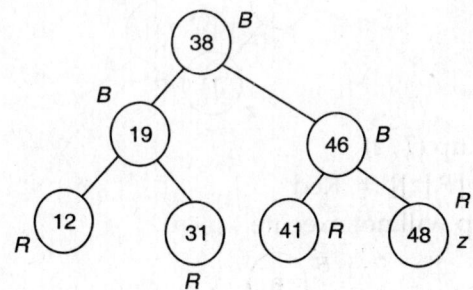

color [P[z]]  ≠  Red
So, while loop terminates

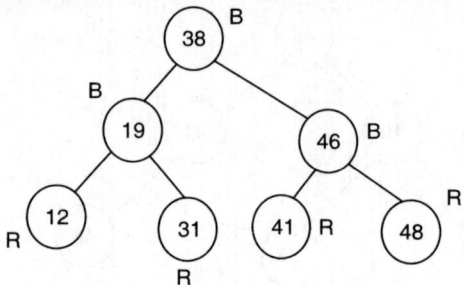

## Example

Show the red-black trees after successively inserting the keys 8, 11, 12, 10, 9, 13 into an initially empty red-black tree.

## Solution

First consider the node with key value 8

RB-Insert $(T, z)$

RB-Insert-Fixup $(T, z)$
        color $[P[z]] \neq$ Red
So, while loop will not execute

Consider the node with key value 11

RB-Insert $(T, z)$

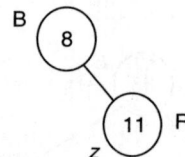

RB-Insert-Fixup $(T, z)$
        color $[P[z]] \neq$ Red
So, while loop will not execute

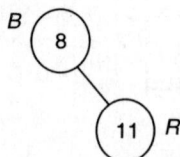

Consider the node with key value 12

RB-Insert $(T, z)$

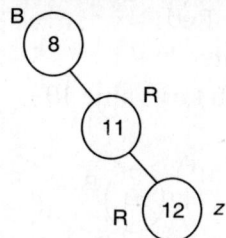

RB-Insert-Fixup $(T, z)$
color $[P[z]] == $ Red
$$P[z] \neq \text{left} [P [P[z]]]$$
$$y = \text{left} [P [P[z]]]$$
So, $y$ points to nil $[T]$

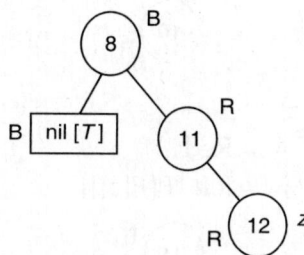

$$\text{color} [y] \neq \text{Red}$$
$$z \neq \text{left} [P[z]]$$
Therefore, case 6 will be applied

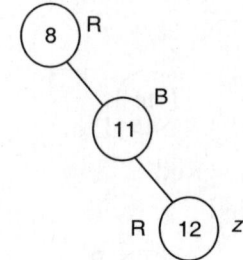

Left-Rotate $(T, P[P[z]])$

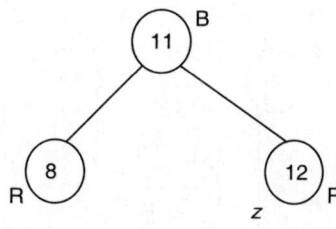

color $[P [z]] \neq$ Red

So, while loop terminates

Consider the node with key value 10

RB-Insert $(T, z)$

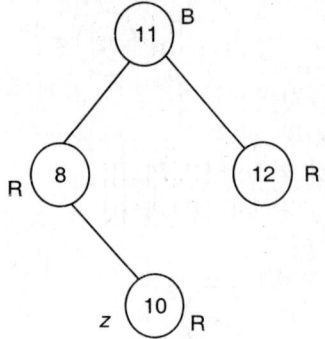

RB-Insert-Fixup $(T, z)$

      color $[P [z]] = =$ Red

        $P [z] = =$ left $[P[P[z]]]$

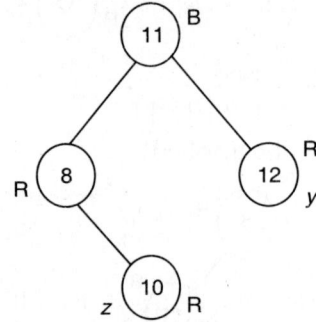

      color $[y] = =$ Red

So, case 1 will be applied

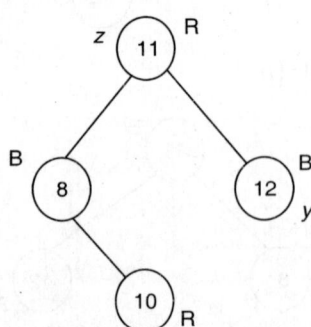

color $[P[z]] \neq$ Red

So, while loop terminates

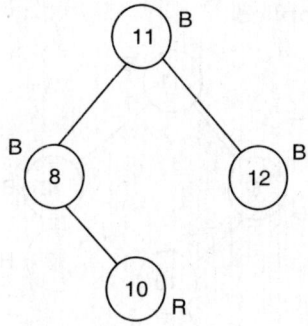

Consider the node with key value 9.

RB-Insert $(T, z)$

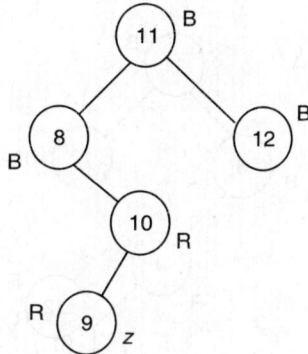

RB-Insert-Fixup $(T, z)$

color $[P[z]] = =$ Red

$P[z] \neq$ left $[P[P[z]]]$

$y$ points to nil $[T]$

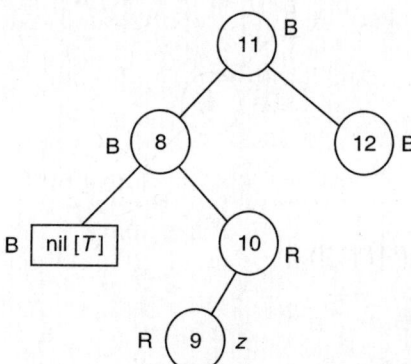

color $[y] \neq$ Red

$z = = $ left $[P [z]]$

So, case 5 will be applied

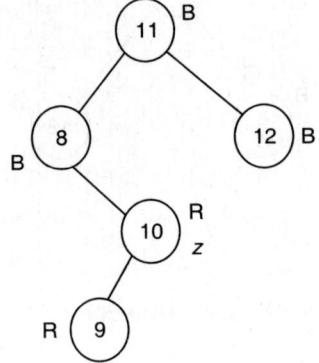

Right-Rotate $(T, z)$

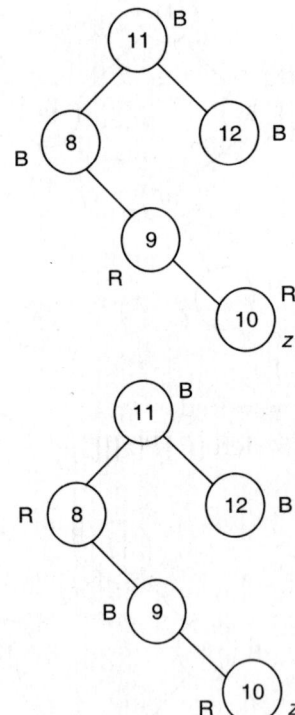

Left-Rotate $(T, P [P[z]])$

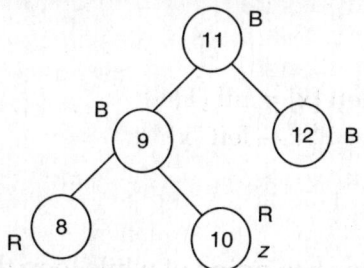

color $[P [z]] \neq$ Red
So, while loop terminates

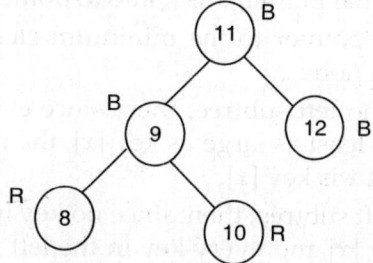

Consider the node with key value 13
RB-Insert $(T, z)$

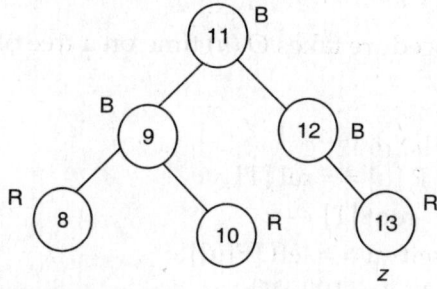

RB-Insert-Fixup $(T, z)$
color $[P [z]] \neq$ Red
So, while loop will not execute

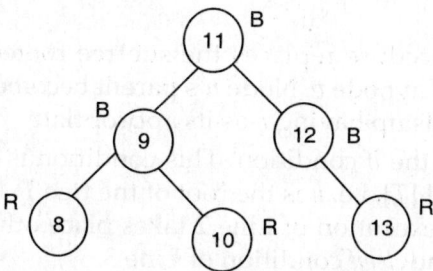

## Algorithm

Tree-minimum (x)

    Line 1     while left [x] ≠ nil [T]

    Line 2             x ← left [x]

    Line 3     return x

## Explanation

- Line 1 indicates the beginning of while loop that ends with Line 2. This loop is applicable till left [x] is not equal to nil [T]. The while loop terminates when left [x] becomes nil [T].
- Line 2 indicates that pointer x is made to point to the left child of x.
- Line 3 returns a pointer to the minimum element in the subtree rooted at a given node x.
- If a node x has no left subtree, then, since every key in the right subtree of x is at least as large as key [x], the minimum key in the subtree rooted at x is key [x].
- If node x has a left subtree, then, since no key in the right subtree is smaller than key [x] and every key in the left subtree is not larger than key [x], the minimum key in the subtree rooted at x resides in the subtree rooted at left [x].

## Analysis

- The above procedure takes $O(h)$ time on a tree of height $h$.

## Algorithm

RB-transplant (T, u, v,)

    Line 1     if  P [u] = = nil [T]

    Line 2        root [T] ← v

    Line 3     elseif  u = = left [P[u]]

    Line 4         left [P [u]] ← v

    Line 5     else   right [P[u]] ← v

    Line 6     P [v] ← P [u]

## Explanation

RB-transplant procedure replaces the subtree rooted at node u with the subtree rooted at node v. Node u's parent becomes node v's parent and u's parent ends up having v as its appropriate child.

- Line 1 checks the *if* condition. This condition is true if the parent of node u is nil [T], i.e. u is the root of the tree T. If this condition is true then the execution of Line 2 takes place otherwise the procedure checks the *elseif* condition of Line 3.

- Line 2 indicates that node $v$ becomes the root of the tree $T$.
- The *elseif* condition of Line 3 is true if the node $u$ is the left child of its parent. If this condition is true then the execution of Line 4 takes place else the control goes to Line 5.
- Line 4 indicates that node $v$ becomes the left child of the parent of $u$.
- Line 5 indicates that node $v$ becomes the right child of the parent of $u$.
- Line 6 indicates that the parent of $u$ becomes the parent of $v$.

### Analysis

- The above procedure takes $O(1)$ time.

### Algorithm

RB-Delete $(T, z)$

| | |
|---|---|
| Line 1 | $y \leftarrow z$ |
| Line 2 | y-original-color $\leftarrow$ color $[y]$ |
| Line 3 | if   left $[z] = = $ nil $[T]$ |
| Line 4 | $x \leftarrow$ right $[z]$ |
| Line 5 | RB-Transplant $(T, z, \text{right } [z])$ |
| Line 6 | elseif   right $[z] = = $ nil $[T]$ |
| Line 7 | $x \leftarrow$ left $[z]$ |
| Line 8 | RB-Transplant $(T, z, \text{left } [z])$ |
| Line 9 | else   $y \leftarrow$ Tree-Minimum (right $[z]$) |
| Line 10 | y-original-color $\leftarrow$ color $[y]$ |
| Line 11 | $x \leftarrow$ right $[y]$ |
| Line 12 | if   P $[y] = = z$ |
| Line 13 | P $[x] \leftarrow y$ |
| Line 14 | else   RB-Transplant $(T, y, \text{right } [y])$ |
| Line 15 | right $[y] \leftarrow$ right $[z]$ |
| Line 16 | P $[\text{right } [y]] \leftarrow y$ |
| Line 17 | RB-Transplant $(T, z, y)$ |
| Line 18 | left $[y] \leftarrow$ left $[z]$ |
| Line 19 | P $[\text{left } [y]] \leftarrow y$ |
| Line 20 | color $[y] \leftarrow$ color $[z]$ |
| Line 21 | if   y-original-color $= = $ Black |
| Line 22 | RB-Delete-Fixup $(T, x)$ |

### Explanation

- Line 1 indicates that pointer $y$ is made to point to node $z$. Node $z$ is that node which we wish to delete from the tree $T$.

- Line 2 indicates that the color of the node pointed by pointer $y$ is stored in a variable $y$-original-color.
- Line 3 checks the *if* condition. This condition is true if $z$ has no left child. If this condition is true then the execution of Lines 4 and 5 takes place else the procedure checks the *elseif* condition of Line 6.
- Line 4 indicates that pointer $x$ is made to point to right child of node $z$.
- Line 5 calls the procedure RB-Transplant $(T, z, \text{right } [z])$.
- Line 6 indicates that the *elseif* condition is true if $z$ has no right child. If this condition is false then the execution of Line 9 takes place.
- Line 7 indicates that the pointer $x$ is made to point to the left child of node $z$.
- Line 8 calls the procedure RB-Transplant $(T, z, \text{left } [z])$.
- Line 9 indicates that pointer $y$ points to the minimum element in the subtree rooted at the right child of node $z$.
- Line 10 indicates that the color of node $y$ is stored in the variable $y$-original-color.
- Line 11 indicates that the pointer $x$ is made to point to the right child of node $y$.
- Line 12 checks the *if* condition. This condition is true if the parent of node $y$ is node $z$. If this condition is true then the execution of Line 13 takes place else the execution of Lines 14, 15 and 16 takes place.
- Line 13 indicates that node $y$ becomes the parent of node $x$.
- Line 14 calls the procedure RB-Transplant $(T, y, \text{right } [y])$.
- Line 15 indicates that the right child of $z$ becomes the right child of $y$.
- Line 16 indicates that node $y$ becomes the parent of the right child of $y$.
- Line 17 calls the procedure RB-Transplant $(T, z, y)$.
- Line 18 indicates that the left child of $z$ becomes the left child of $y$.
- Line 19 indicates that node $y$ becomes the parent of the left child of $y$.
- Line 20 indicates that the color of node $z$ is assigned to node $y$.
- Line 21 checks the 'if' condition. This condition is true if the variable $y$-original-color stores black color. If this condition is true then the execution of Line 22 takes place.
- Line 22 calls the procedure RB-Delete-Fixup $(T, x)$.

## Algorithm

RB-Delete-Fixup (T, x)

| | | |
|---|---|---|
| | Line 1 | while  $x \neq$ root [T] and color [x] = = Black |
| | Line 2 | if   $x ==$ left [P [x]] |
| | Line 3 | $w \leftarrow$ right [P [x]] |
| | Line 4 | if  color [w] = = Red |
| Case 1 | Line 5 | color [w] $\leftarrow$ Black |
| | Line 6 | color [P[x]] $\leftarrow$ Red |
| | Line 7 | Left-Rotate (T, P [x]) |
| | Line 8 | $w \leftarrow$ right [P [x]] |
| | Line 9 | if  color [left [w]] = = Black and color [right [w]] = = Black |
| Case 2 | Line 10 | color [w] $\leftarrow$ Red |
| | Line 11 | $x \leftarrow$ P [x] |
| | Line 12 | elseif  color [right [w]] = = Black |
| Case 3 | Line 13 | color [left [w]] $\leftarrow$ Black |
| | Line 14 | color [w] $\leftarrow$ Red |
| | Line 15 | Right-Rotate (T, w) |
| | Line 16 | $w \leftarrow$ right [P [x]] |
| | Line 17 | color [w] $\leftarrow$ color [P[x]] |
| Case 4 | Line 18 | color [P [x]] $\leftarrow$ Black |
| | Line 19 | color [right [w]] $\leftarrow$ Black |
| | Line 20 | Left-Rotate (T, P [x]) |
| | Line 21 | $x \leftarrow$ root [T] |
| | Line 22 | else   $w \leftarrow$ left [P [x]] |
| | Line 23 | if  color [w] = = Red |
| | Line 24 | color [w] $\leftarrow$ Black |
| Case 5 | Line 25 | color [P [x]] $\leftarrow$ Red |
| | Line 26 | Right-Rotate (T, P [x]) |
| | Line 27 | $w \leftarrow$ left [P [x]] |
| | Line 28 | if  color [right [w]] = = Black and color [ left [w]] = = Black |
| Case 6 | Line 29 | color [w] $\leftarrow$ Red |
| | Line 30 | $x \leftarrow$ P [x] |
| | Line 31 | elseif   color [left [w]] = = Black |
| Case 7 | Line 32 | color [right [w]] $\leftarrow$ Black |
| | Line 33 | color [w] $\leftarrow$ Red |
| | Line 34 | Left-Rotate (T, w) |
| | Line 35 | $w \leftarrow$ left [P [x]] |

Case 8 $\begin{cases} \text{Line 36} \\ \text{Line 37} \\ \text{Line 38} \\ \text{Line 39} \\ \text{Line 40} \end{cases}$ $\quad$ color [w] ← color [P [x]]
$\qquad\qquad\qquad\quad$ color [P [x]] ← Black
$\qquad\qquad\qquad\quad$ color [left [w]] ← Black
$\qquad\qquad\qquad\quad$ Right-Rotate (T, P [x])
$\qquad\qquad\qquad\quad$ x ← root [T]

$\qquad$ Line 41 $\quad$ color [x] ← Black

## Explanation

- Line 1 indicates the beginning of while loop that ends with Line 40. This loop is applicable until $x \neq$ root $[T]$ and color of node $x$ is black.
- Line 2 checks the *if* condition. This condition is true if $x$ is the left child of its parent. If this condition is false then the control goes to Line 22.
- Line 3 indicates that pointer $w$ is made to point to the right child of the parent of $x$, or in other words $w$ points to the sibling of $x$.
- Line 4 checks the 'if' condition. This condition is true if the color of node $w$ is red. If this condition is true then the execution of Line 5 to Line 8 takes place otherwise the procedure checks the 'if' condition of Line 9.
- Line 5 assigns black color to node $w$.
- Line 6 assigns red color to the parent of node $x$.
- Line 7 calls the procedure Left-Rotate $(T, P [x])$.
- Line 8 indicates that the pointer $w$ is made to point to the right child of the parent of $x$.
- Line 9 checks the *if* condition. This condition is true if color [left [$w$]] is black and color [right [$w$]] is black. If this condition is true then the execution of Lines 10 and 11 takes place otherwise the control goes to Line 12.
- Line 10 indicates that red color is assigned to node $w$ (sibling of $x$).
- Line 11 indicates that pointer $x$ is made to point to the parent of $x$.
- Line 12 checks the *elseif* condition. This condition is true if color [right [$w$]] is black. If this condition is true then the execution of Line 13 to 16 takes place. If this condition is false then the control goes to Line 17.
- Line 13 indicates that a black color is assigned to the left child of $w$.
- Line 14 indicates that a red color is assigned to the node $w$.

- Line 15 calls the procedure Right-Rotate $(T, w)$.
- Line 16 indicates that pointer $w$ is made to point to the right child of the parent of $x$.
- Line 17 assigns the same color to node $w$ as that of parent of $x$.
- Line 18 assigns black color to the parent of $x$.
- Line 19 assigns black color to the right child of node $w$.
- Line 20 calls the procedure Left-Rotate $(T, P[x])$.
- Line 21 indicates that pointer $x$ is made to point to the root of tree $T$.
- Line 22 indicates that pointer $w$ is made to point to the left child of the parent of $x$ or in other words $w$ points to the sibling of $x$.
- Line 23 checks the *if* condition. This condition is true if the color of node $w$ is red. If this condition is true then the execution of Line 24 to Line 27 takes place otherwise the procedure checks the *if* condition of Line 28.
- Line 24 assigns black color to node $w$.
- Line 25 assigns red color to the parent of node $x$.
- Line 26 calls the procedure Right-Rotate $(T, P[x])$.
- Line 27 indicates that the pointer $w$ is made to point to the left child of the parent of $x$.
- Line 28 checks the *if* condition. This condition is true if color [right [$w$]] is black and color [left [$w$]] is black. If this condition is true then the execution of Lines 29 and 30 takes place otherwise the control goes to Line 31.
- Line 29 indicates that red color is assigned to node $w$.
- Line 30 indicates that pointer $x$ is made to point to the parent of $x$.
- Line 31 checks the *elseif* condition. This condition is true if color [left [$w$]] is black. If this condition is true then the execution of Line 32 to 35 takes place. If this condition is false then the control goes to Line 36.
- Line 32 indicates that a black color is assigned to the right child of $w$.
- Line 33 indicates that a red color is assigned to node $w$.
- Line 34 calls the procedure Left-Rotate $(T, w)$.
- Line 35 indicates that pointer $w$ is made to point to the left child of the parent of $x$.
- Line 36 assigns the same color to node $w$ as that of parent of $x$.
- Line 37 assigns black color to the parent of $x$.
- Line 38 assigns black color to the left child of node $w$.
- Line 39 calls the procedure Right-Rotate $(T, P[x])$.

- Line 40 indicates that pointer $x$ is made to point to the root of tree $T$.
- Line 41 assigns black color to node $x$.

❑ During the execution of the procedure RB-Delete $(T, z)$ red-black properties 1, 2 and 4 may be violated. To restore these red black properties, RB-Delete-Fixup $(T, x)$ procedure is called.

Property 1 says that every node in a RB tree is either red or black.

Property 2 says that the root of RB tree is black.

Property 4 says that if a node is red, then both its children are black.

The violation of these properties occurs if node $y$ was black in color.

When we remove or move the black node $y$, we push its blackness onto node $x$. This causes node $x$ to be either 'doubly black' or 'red and black', thereby violating property 1. Remember if a node $x$ is doubly black it contributes 2 to the count of black nodes on simple paths containing $x$. If a node $x$ is red and black, it contributes 1 to the count of black nodes on simple paths containing $x$.

Property 2 is violated if $y$ had been the root and a red child of $y$ becomes the new root.

Property 4 is violated if both $x$ and $P[x]$ are red.

❑ For better understanding of the pseudo code of RB-Delete-Fixup $(T, x)$, it is divided into eight cases. Cases 5, 6, 7, 8 are same as cases 1, 2, 3, 4 respectively with 'left' and 'right' exchanged. Once we enter into case 3, then we have to enter into case 4. So we can enter into case 4 either directly or through case 3. Similarly, once we enter into case 7, then we have to enter into case 8. So we can enter into case 8 either directly or through case 7.

❑ If $x$ is the left child of its parent, then case 1, case 2, case 3 and case 4 are taken into account.

❑ If $x$ is the right child of its parent, then case 5, case 6, case 7 and case 8 are taken into account.

❑ Case 1 occurs if $x$'s sibling $w$ is red. Line 5 to Line 8 constitutes case 1.

❑ Case 2 occurs if $x$'s sibling $w$ is black, and both of $w$'s children are black.

Line 10 and Line 11 constitute case 2.

❑ Case 3 occurs if $x$'s sibling $w$ is black, $w$'s left child is red, and $w$'s right child is black. Line 13 to Line 16 constitutes case 3.

❑ Case 4 occurs if $x$'s sibling $w$ is black, and $w$'s right child is red.

Line 17 to Line 21 constitutes case 4.

❑ Case 5 occurs if $x$'s sibling $w$ is red.

❑   Line 24 to Line 27 constitutes case 5.
❑   Case 6 occurs if $x$'s sibling $w$ is black, and both of $w$'s children are black.

Line 29 and Line 30 constitute case 6.

❑   Case 7 occurs if $x$'s sibling $w$ is black, $w$'s right child is red and $w$'s left child is black. Line 32 to Line 35 constitutes case 7.
❑   Case 8 occurs if $x$'s sibling $w$ is black, and $w$'s left child is red. Line 36 to Line 40 constitutes case 8.

### Analysis

- Since the height of RB-tree of $n$ nodes is $O\ (\lg n)$, Lines 1-21 of RB-Delete take $O\ (\lg n)$ time.
- In RB-Delete-Fixup, each of cases 1, 3 and 4 lead to termination after performing a constant number of color changes and at most three rotations.
- Case 2 is the only case in which the while loop can be repeated, and then the pointer $x$ moves up the tree at most $O\ (\lg n)$ times, performing no rotations. Thus the procedure RB-Delete-Fixup takes $O\ (\lg n)$ time and performs at most three rotations.
- The overall time for RB-Delete is therefore $O\ (\lg n)$.

### Example

Show the red black trees that result from the successive deletion of the keys in the order 46, 48, 19, 12, 31, 38, 41.

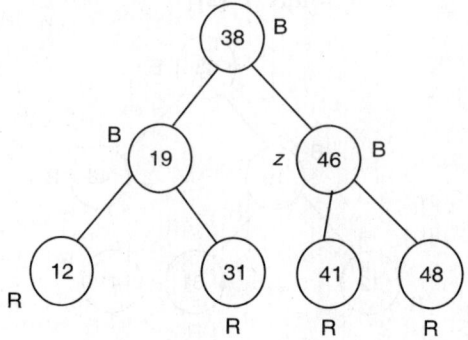

### Solution

Delete node with key value 46

RB-Delete $(T, z)$

$y$ points to node with key 46

   $y$-original-color  =  Black

           left $[z]$  ≠  nil $[T]$

       right $[z]$  ≠  nil $[T]$

Call Tree-Minimum (right [z])

Since left [right [z]] = = nil [T]

while loop of Tree-Minimum procedure will not execute

Hence the procedure returns node 48

So, y points to node 48

    y-original-color = Red

x points to nil [T]

$$P [y] = = z$$

node 48 becomes the parent of nil [T]

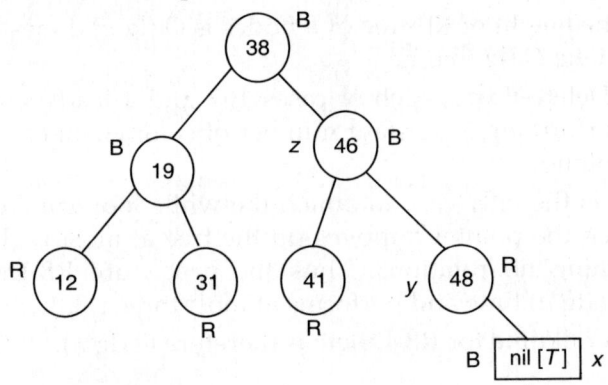

RB-Transplant (T, z, y)

$$P [z] \neq nil [T]$$

$$z = = left [P [z]]$$

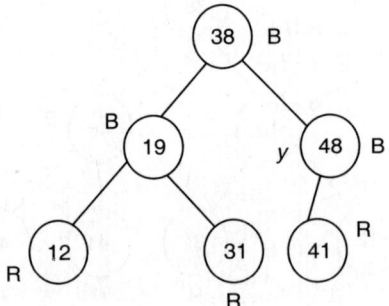

    y-original-color ≠ Black

Delete node with key value 48

RB-Delete (T, z)

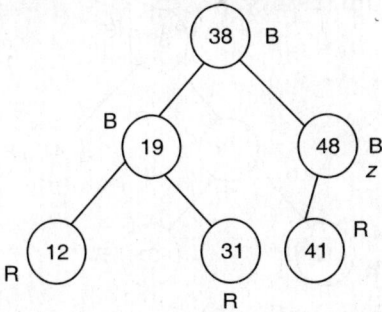

*y* points to node with key value 48

    *y*-original-color = Black

        left [*z*] ≠ nil [*T*]

        right [*z*] = = nil [*T*]

pointer *x* points to node 41

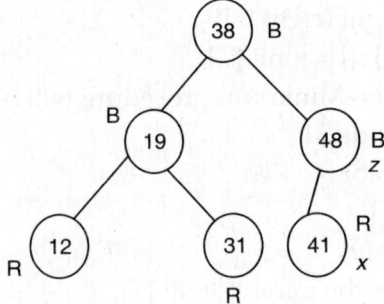

RB-Transplant (*T*, *z*, left [*z*])

        $P[z]$ ≠ nil [*T*]

        *z* ≠ left [*P* [*z*]]

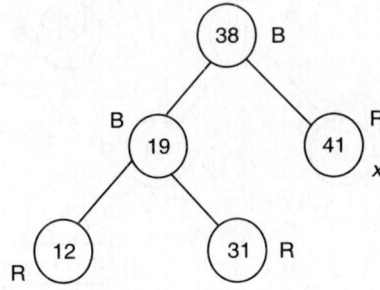

    *y*-original-color = Black

RB-Delete-Fixup (*T*, *x*)

        *x* ≠ root [*T*] and color [*x*] ≠ Black

So, while loop will not execute.

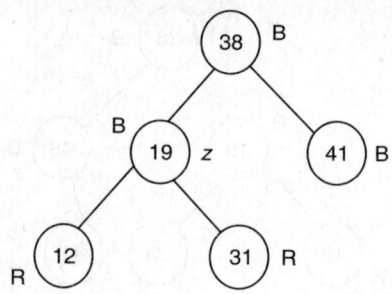

Delete node with key value 19.

RB-Delete $(T, z)$

$y$ points to node with key 19

    $y$-original-color = Black

        left $[z] \neq$ nil $[T]$

       right $[z] \neq$ nil $[T]$

call Tree-Minimum (right $[z]$)

since left [right $[z]$] = = nil $[T]$,

while loop of Tree-Minimum procedure will not execute.

So, $y$ points to node 31

    $y$-original-color = Red

$x$ points to nil $[T]$

        $P[y]$ = = $z$

node 31 becomes the parent of nil $[T]$.

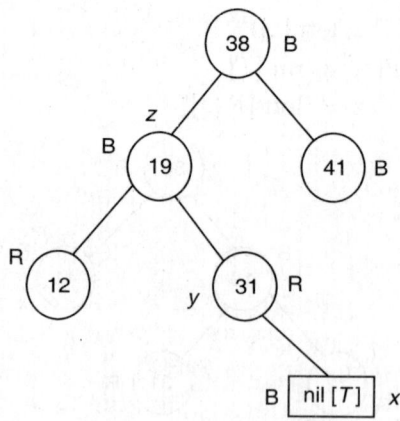

RB-Transplant $(T, z, y)$

        $P[z] \neq$ nil $[T]$

          $z$ = = left $[P[z]]$

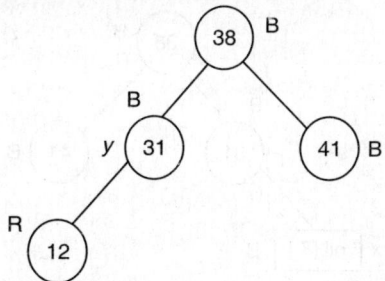

$y$-original-color ≠ Black

Delete node with key value 12

RB-Delete ($T$, $z$)

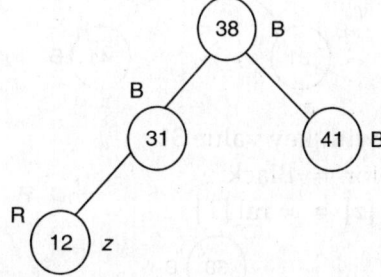

$y$ points to node with key value  12

    $y$-original-color = Red

        left [$z$] = = nil [$T$]

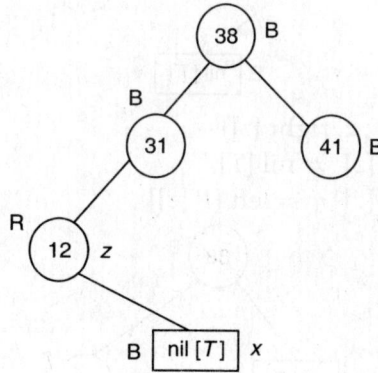

RB-Transplant ($T$, $z$, right [$z$])

      $P$ [$z$] ≠ nil [$T$]

        $z$ = = left [$P$ [$z$]]

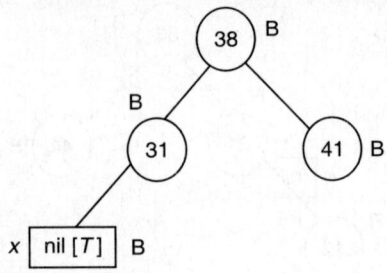

y-original-color ≠ Black
Delete node with key value 31

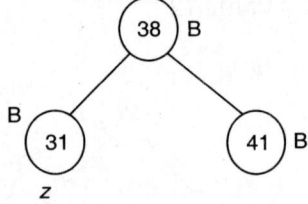

$y$ points to node with key value 31
    y-original-color = Black
        left $[z]$ = = nil $[T]$

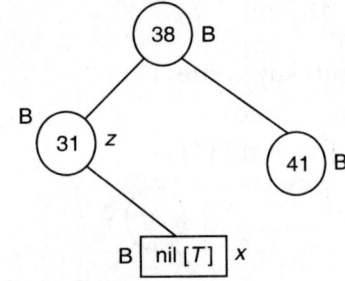

RB-Transplant $(T, z, \text{right} [z])$
        $P [z] \neq$ nil $[T]$
        $P [z] = =$ left $[P [z]]$

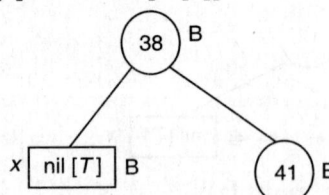

    y-original-color = = Black
RB-Delete-Fixup $(T, x)$
            $x \neq$ root $[T]$ and color $[x] = =$ Black
            $x = =$ left $[P [[x]]]$

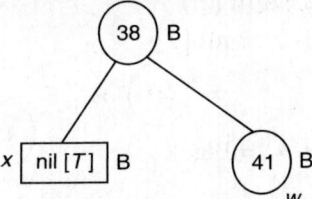

color $[w] \neq$ Red

color [left $[w]$] == Black and color [right $[w]$] == Black

So, case 2 will be applied

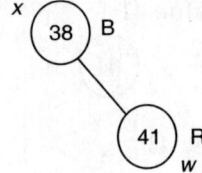

$$x == \text{root } [T],$$

So, while loop terminates

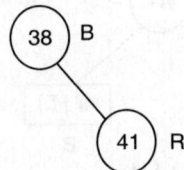

Delete node with key value 38

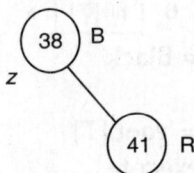

RB-Delete $(T, z)$

$y$ points to node with key 38

    $y$-original-color = Black

        left $[z]$ == nil $[T]$

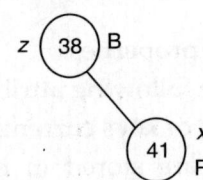

RB-Transplant $(T, z, \text{right }[z])$

$$P[z] = = \text{nil }[T]$$

$x$  R

$y$-original-color $= =$ Black

RB-Delete-Fixup $(T, x)$

$$x = = \text{root }[T]$$

So, while loop will not execute

 B

Delete node with key value 41

$z$ $\boxed{41}$ B

RB-Delete $(T, z)$

$y$ points to node with key 41

$y$-original-color $=$ Black

$$\text{left }[z] = = \text{nil }[T]$$

$\boxed{41}$ B

$z$

$\boxed{\text{nil }[T]}$ $x$

B

RB-Transplant $(T, z, \text{right }[z])$

$$P[z] = = \text{nil }[T]$$

B $\boxed{\text{nil }[T]}$ $x$

$y$-original-color $= =$ Black

RB-Delete-Fixup $(T, x)$

$$x = = \text{root }[T]$$

So, while loop will not execute

B $\boxed{\text{nil }[T]}$

## 2.6 B-TREES

A B-tree is similar to RB tree with the only difference that a B-tree may have many children.

A B-tree has the following properties:

1. Every node $x$ has the following attributes:
   i. $n[x]$, the number of keys currently stored in node $x$.
   ii. The $n[x]$ keys are stored in non-decreasing order, i.e. $\text{key}_1[x] \le \text{key}_2[x] \le \ldots \le \text{key}_{n[x]}[x]$.

iii. leaf $[x]$, a Boolean value which is true, if $x$ is a leaf and false, if $x$ is an internal node.

2. Each internal node $x$ also contains $n[x] + 1$ pointers, $C_1 [x]$, $C_2 [x]$, ..., $C_{n+1[x]} [x]$ to its children.

3. The keys, $key_i [x]$ separate the ranges of keys stored in each subtree. If $k_i$ is any key stored in the subtree with root $C_i [x]$, then

$$k_i \leq key_1 [x] \leq k_2 \leq key_2 [x] \leq \dots \leq key_{n[x]} [x] \leq k_{n+1[x]}.$$

4. All leaves have the same depth, which is the tree's height $h$.

   If $n \geq 1$, then for any $n$-key B-tree T of height $h$ and minimum degree $t \geq 2$, $h \leq \log_t \dfrac{n+1}{2}$

5. Nodes in a B-tree have lower and upper bounds on the number of keys they can contain. These bounds are expressed in terms of a fixed integer $t \geq 2$ called the minimum degree of the B-tree.

   i. Every node other than the root must have at least $t - 1$ keys. Every internal node other than the root thus has at least $t$ children. If the tree is non empty, the root must have at least one key.

   ii. Every node may contain at most $2t - 1$ keys. Therefore an internal node may have at most $2t$ children. A node is said to be 'full' if it contains exactly $2t - 1$ keys.

   Remember that we cannot insert a key into a leaf node which is full.

   Before inserting a key we have to 'split' that full node.

- In a B-tree application, the amount of data handled is so large that all the data do not fit into the main memory at once. So, the B-tree algorithms copy selected pages from disk into main memory as needed and write back onto disk the pages that have changed. B-tree algorithms keep only a constant number of pages in main memory at any time.

- Let $x$ be a pointer to an object. If the object is currently in the computer's main memory, then we can refer to the attributes of the object as usual.

  If the object referred to by $x$ resides on disk, then an operation Disk-Read $(x)$ is performed to read object $x$ into main memory before referring to its attributes. Similarly, the operation Disk-Write $(x)$ is used to save any changes that have been made to the attributes of object $x$.

## Algorithm

B-Tree-Search $(x, k)$

> Line 1      $i \leftarrow 1$
>
> Line 2      while  $i \leq n[x]$ and $k > key_i[x]$
>
> Line 3           $i \leftarrow i + 1$
>
> Line 4      if   $i \leq n[x]$ and $k == key_i[x]$
>
> Line 5           return $(x, i)$
>
> Line 6      elseif  leaf $[x]$
>
> Line 7           return nil
>
> Line 8      else   Disk-Read $(C_i[x])$
>
> Line 9           return   B-Tree-Search $(C_i[x], k)$

## Explanation

- The above algorithm B-Tree-Search $(x, k)$ takes as input a pointer to the root node $x$ of a subtree and a key $k$ to be searched for in that subtree. The first call is of the form B-Tree-Search (root $[T], k$).
- Line 1 indicates that the index $i$ is set to 1.
- Line 2 indicates the beginning of while loop that ends with Line 3. This loop is applicable until $i$ is less than or equal to the number of keys currently stored in node $x$ and key $k$ is greater than the $key_i[x]$.
- Line 3 indicates that the value of $i$ is incremented by 1.
- Line 4 checks the *if* condition. This condition is true, if index $i$ is less than or equal to the number of keys currently stored in node $x$ and key $k$ is equal to $key_i[x]$. If this condition is true then the execution of Line 5 takes place else the control goes to Line 6.
- Line 5 returns node $x$ and an index $i$ such that $key_i[x] == k$.
- Line 6 checks the *elseif* condition. This condition is true if $x$ is a leaf. If this condition is true then the execution of Line 7 takes place else the control goes to Line 8.
- Line 7 returns nil and the search terminates.
- Line 8 calls the Disk-Read $(C_i[x])$ operation.
- Line 9 recurses the procedure to search the appropriate subtree of $x$ by calling B-Tree-Search $(C_i[x], k)$.

## Analysis

- The B-Tree-Search procedure accesses $O(h) = O(\log_t n)$ disk pages, where $h$ is the height of the B-tree and $n$ is the number of keys in the B-tree.
- Since $n[x] < 2t$, the while loop of Lines 2 and 3 takes $O(t)$ time within each node and the total CPU time is $O(th) = O(t \log_t n)$.

## Example

Search key $S$ in the following B-Tree:

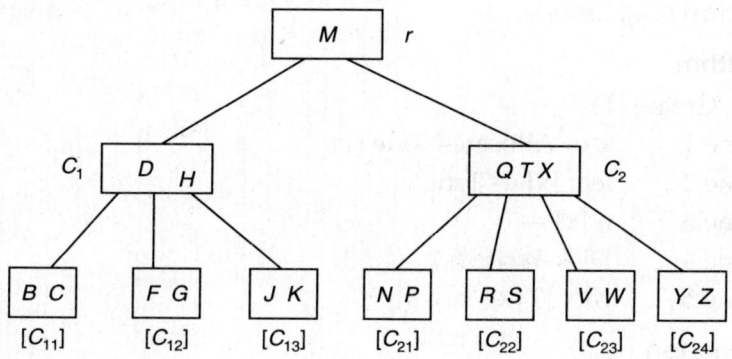

## Solution

B-Tree-Search $(r, S)$

$$i = 1$$
$$1 \leq 1 \quad \text{and} \quad S > M$$
$$i = 2$$
$$2 \nleq 1$$

So, while loop terminates.

$$2 \nleq 1$$

*if* condition fails.

Pointer is pointing towards root,

So, *elseif* condition fails

Disk-Read $(C_2)$

B-Tree-Search $(C_2, S)$

$$i = 1$$
$$1 \leq 3 \text{ and } S > Q$$
$$i = 2$$
$$2 \leq 3 \text{ and } S \ngtr T$$

So, while loop terminates.

$$2 \leq 3 \text{ and } S \neq T$$

*if* condition fails

Pointer is pointing towards the internal node $C_2$

So, *elseif* condition fails

Disc-Read $(C_{22})$

B-Tree-Search $(C_{22}, S)$

$$i = 1$$
$$1 \leq 2 \quad \text{and} \quad S > R$$
$$i = 2$$
$$2 \leq 2 \quad \text{and} \quad S \ngtr S$$

So, while loop terminates

$$2 \leq 2 \quad \text{and} \quad S == S$$

return $(C_{22}, 2)$

## Algorithm

B-Tree-Create (T)

| | |
|---|---|
| Line 1 | $x \leftarrow$ Allocate-Node ( ) |
| Line 2 | leaf [x] $\leftarrow$ True |
| Line 3 | n [x] $\leftarrow$ 0 |
| Line 4 | Disk-Write (x) |
| Line 5 | root [T] $\leftarrow$ x |

## Explanation

- The procedure Allocate-Node ( ) allocates one disk page to be used as a new node.
- Line 1 indicates that $x$ is the new node created by Allocate-Node ( ) procedure.
- Line 2 sets leaf [$x$] as True which means that $x$ is a leaf.
- Line 3 sets $n[x]$ as 0 which means that node $x$ is empty.
- Line 4 calls the procedure Disk-Write ($x$).
- Line 5 indicates that node $x$ becomes the root of the B-Tree T.
- To build a B-tree T, first use B-Tree-Create ($T$) procedure to create an empty root node and then call B-Tree-Insert to add new keys.

## Analysis

- B-Tree-Create procedure needs $O$ (1) disk operations and $O$ (1) CPU time.

## Algorithm

B-Tree-Insert (T, k)

| | | |
|---|---|---|
| Line 1 | | $r \leftarrow$ root [T] |
| Line 2 | | if n [r] $== 2t - 1$ |
| Line 3 | | $s \leftarrow$ Allocate-Node ( ) |
| Line 4 | | root [T] $\leftarrow$ s |
| Line 5 | | leaf [s] $\leftarrow$ False |
| Line 6 | | n [s] $\leftarrow$ 0 |
| Line 7 | | $C_1$ [s] $\leftarrow$ r |
| Line 8 | | B-Tree-Split-Child (s, 1) |
| Line 9 | | B-Tree-Insert-Nonfull (s, k) |
| Line 10 | else | B-Tree-Insert-Nonfull (r, k) |

## Explanation

- This procedure inserts a key into a B-tree.
- Line 1 indicates that pointer $r$ is pointing towards the root of the B-Tree $T$.
- Line 2 checks the *if* condition. If this condition is true then the execution of Lines 3 to 9 takes place else the control goes to Line 10.

  This condition is true if the root node $r$ is currently having $2t - 1$ keys. In other words root node $r$ is full.
- Line 3 indicates that node $s$ is the new node created by Allocate-Node ( ) procedure.
- Line 4 indicates the node $s$ becomes the new root.
- Line 5 indicates that $s$ is an internal node.
- Line 6 indicates that node $s$ is an empty node.
- Line 7 indicates that node $r$ becomes the child of node $s$.
- Line 8 calls the procedure B-Tree-Split-child $(s, 1)$.
- Line 9 calls the procedure B-Tree-Insert-Nonfull $(s, k)$.
- Line 10 calls the procedure B-Tree-Insert-Nonfull $(r, k)$.

## Analysis

- A key '$k$' is inserted into a B-tree $T$ of height $h$ in a single pass down the tree, requiring $O(h)$ disk accesses because only $O(1)$ Disk-Read and Disk-Write operations occur between calls to B-Tree-Insert-Nonfull.
- The CPU time required is $O(th) = O(t \log_t n)$.
- B-Tree-Insert-Nonfull is tail recursive, therefore, it can be alternatively implemented as a while loop. So, the number of pages that need to be in main memory at any time is $O(1)$.

## Algorithm

B-Tree-Split-Child $(x, i)$

| | |
|---|---|
| Line 1 | $z \leftarrow$ Allocate-Node ( ) |
| Line 2 | $y \leftarrow C_i [x]$ |
| Line 3 | leaf $[z] \leftarrow$ leaf $[y]$ |
| Line 4 | $n [z] \leftarrow t - 1$ |
| Line 5 | for $j \leftarrow 1$ to $t - 1$ |
| Line 6 | $key_j [z] \leftarrow key_{j + t} [y]$ |
| Line 7 | if not leaf $[y]$ |
| Line 8 | for $j \leftarrow 1$ to $t$ |
| Line 9 | $C_j [z] \leftarrow C_{j + t} [y]$ |
| Line 10 | $n [y] \leftarrow t - 1$ |

Line 11        for  $j \leftarrow n[x] +1$ down to $i + 1$

Line 12              $C_{j+1}[x] \leftarrow C_j[x]$

Line 13        $C_{i+1}[x] \leftarrow z$

Line 14        for  $j \leftarrow n[x]$ down to i

Line 15              $key_{j+1}[x] \leftarrow key_j[x]$

Line 16        $key_i[x] \leftarrow key_t[y]$

Line 17        $n[x] \leftarrow n[x] +1$

Line 18        Disk-Write (y)

Line 19        Disk-Write (z)

Line 20        Disk-Write (x)

## Explanation

- The above procedure B-Tree-Split-Child takes as input a nonfull internal node $x$ and an index $i$ such that $C_i[x]$ is a full child of $x$. Both $x$ and $C_i[x]$ are assumed to be in main memory.
- The procedure splits this full child $C_i[x]$ about its median key $S$, which moves up into $C_i[x]$'s parent node $x$.
- The keys in $C_i[x]$ that are greater than the median key move into a new node $z$, which becomes a new child of $x$.
- Line 1 indicates that $z$ is a new node created by the procedure Allocate-Node ( ).
- Line 2 indicates $y$ is $x$'s $i$th child.
- Line 3 indicates that node $z$ is a leaf node.
- Line 4 indicates that $t - 1$ keys should be stored in node $z$.
- Line 5 indicates the beginning of for loop that ends with Line 6. This for loop is applicable for $j$ equals to 1 to $t - 1$.
- Line 6 indicates that the keys in $y$ larger than median key move to the node $z$.
- Line 7 checks the *if* condition. This condition is true, if $y$ is not a leaf. If this condition is false then the control goes to Line 10.
- Line 8 indicates the beginning of for loop that ends with Line 9. This loop is applicable for $j$ equals to 1 to $t$.
- Line 9 indicates that the children of node $y$ become the children of new node $z$.
- Line 10 indicates that $t - 1$ keys should be stored in node $y$. Originally node $y$ has $2t - 1$ keys but after splitting node $y$, this number should reduces to $t - 1$.
- Line 11 indicates the beginning of for loop that ends with Line 12. Index $j$ runs from $n[x] +1$ down to $i + 1$.

- Line 12 indicates $j$th child of $x$ moves one position ahead.
- Line 13 indicates that node $z$ becomes the $(i + 1)$th child of $x$.
- Line 14 indicates the beginning of for loop that ends with Line 15. Index $j$ runs from $n[x]$ down to $i$.
- Line 15 indicates that key of $x$ at $j$th location moves one position ahead.
- Line 16 indicates that the key of $y$ at $(t)$ th location becomes the key of $x$ at ith location. In this way the median key moves from $y$ up to $x$.
- Line 17 indicates that the number of keys stored in $x$ has been incremented by 1. It happens because the median key of $y$ moves up to become the key in $x$.
- Line 18 calls the Disk-Write $(y)$ procedure.
- Line 19 calls the Disk-Write $(z)$ procedure.
- Line 20 calls the Disk-Write $(x)$ procedure and all the above three calls write the modified disk pages.

## Analysis

- The loops of Lines 5-6 and 8-9 run for $\Theta(t)$ iterations.
  The other loops run for $O(t)$ iterations. So the CPU time used by B-Tree- Split-Child is $\Theta(t)$.
- B-Tree-Split-Child performs $O(1)$ disk operations.

## Algorithm

B-Tree-Insert-Nonfull $(x, k)$

| | | |
|---|---|---|
| Line 1 | | $i \leftarrow n[x]$ |
| Line 2 | | if leaf $[x]$ |
| Line 3 | | while $i \geq 1$ and $k < key_i[x]$ |
| Line 4 | | $key_{i+1}[x] \leftarrow key_i[x]$ |
| Line 5 | | $i \leftarrow i - 1$ |
| Line 6 | | $key_{i+1}[x] \leftarrow k$ |
| Line 7 | | $n[x] \leftarrow n[x] + 1$ |
| Line 8 | | Disk-Write $(x)$ |
| Line 9 | else | while $i \geq 1$ and $k < key_i[x]$ |
| Line 10 | | $i \leftarrow i - 1$ |
| Line 11 | | $i \leftarrow i + 1$ |
| Line 12 | | Disk-Read $(C_i[x])$ |
| Line 13 | | if $n[C_i[x]] == 2t - 1$ |
| Line 14 | | B-Tree-Split-Child $(x, i)$ |

Line 15                        if   k > key$_i$ [x]
Line 16                            i ← i + 1
Line 17                    B-Tree-Insert-Nonfull (C$_i$ [x], k)

## Explanation

- This procedure inserts key $k$ into node $x$, which is assumed to be non-full when the procedure is called.
- Line 1 indicates that $i$ stores the number of keys currently present in node $x$.
- Line 2 checks the *if* condition. This condition is true when $x$ is a leaf node. If this condition is false then the control goes to Line 9.
- Line 3 indicates the beginning of while loop that ends with Line 5. This loop is applicable for $i \geq 1$ and $k < $ key$_i$ [x].
- Line 4 indicates that the key of $x$ at $i$th location moves one position ahead.
- Line 5 indicates that the value of $i$ is decremented by 1.
- Line 6 indicates that key $k$ is inserted in $x$ at $(i + 1)$th location.
- Line 7 indicates that the number of keys in $x$ is incremented by 1. It happens because of the insertion of new key $k$.
- Line 8 calls the Disk-Write $(x)$ operation.
- Line 9 indicates the beginning of while loop that ends with Line 10. This loop is applicable for $i \geq 1$ and $k < $ key$_i$ [x].
- Line 10 indicates that the value of $i$ is decremented by 1.
- Line 11 indicates that the value of $i$ is incremented by 1.
- Lines 9 to 11 determine the child of $x$ to which the recursion descends.
- Line 12 calls the procedure Disk-Read (C$_i$ [x]).
- Line 13 checks the *if* condition. This condition is true if number of keys in C$_i$ [x] is equal to $2t - 1$. It means that the recursion descends to a full child. If this condition is false then the control goes to Line 17.
- Line 14 calls the B-Tree-Split-Child $(x, i)$ procedure to split that child into two non-full children.
- Line 15 checks the *if* condition. This condition is true when $k$ is greater than key$_i$ [x].
- Line 16 indicates that the value of i is incremented by 1.
- Lines 15 to 16 determine which of the two children is now the correct one to descend to.

- The execution of Line 13 to 16 takes place to ensure that the procedure never recurses to a full node.
- Line 17 calls the procedure B-Tree-Insert-Non-full ($C_i$ [$x$], $k$) to insert $k$ into an appropriate subtree of $x$.

## Example

Insert the following keys into an empty B-tree with minimum degree 2
$B, D, G, H, J, K, M, O, P, R$

## Solution

To build a B-Tree $T$, we first call B-Tree-Create procedure to create an empty root node and then call B-Tree-Insert to add new keys.

B-Tree-Insert ($T, B$)

$$n [r] \neq 2t - 1 \quad (2t - 1 = 2 \times 2 - 1 = 3)$$

B-Tree-Insert-Nonfull ($r, B$)

$$i = 0$$

$r$ is a leaf

$$i \not\geq 1$$

So, while loop will not execute

$$\boxed{B} \;\; r$$

$$n [r] = 1$$

B-Tree-Insert ($T, D$)

$$n [r] \neq 3$$

B-Tree-Insert-Nonfull ($r, D$)

$$i = 1$$

$r$ is a leaf

$$1 \geq 1 \text{ and } D \not< B$$

So, while loop will not execute

$$\boxed{B \; D} \;\; r$$

$$n [r] = 2$$

B-Tree-Insert ($T, G$)

$$n [r] \neq 3$$

B-Tree-Insert-Nonfull ($r, G$)

$$i = 2$$

$r$ is a leaf

$$2 \geq 1 \text{ and } G \not< D$$

So, while loop will not execute

$$\boxed{B \; D \; G} \;\; r$$

$$n\ [r]\ =\ 3$$

B-Tree-Insert $(T, H)$

$$n\ [r]\ =\ =\ 3$$

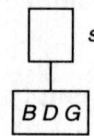

B-Tree-Split-Child $(s, 1)$

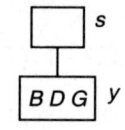

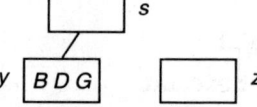

$n\ [z]$ should be 1

$$j\ =\ 1$$

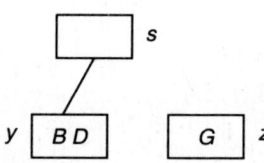

*if* condition fails because $y$ is a leaf node

$$n\ [y]\ =\ 1$$

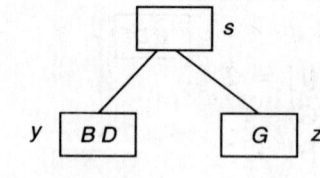

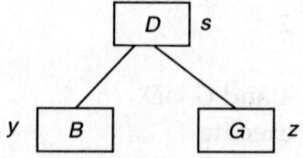

$$n\,[s] \;=\; 1$$

B-Tree-Insert-Nonfull $(s, H)$

$$i \;=\; 1$$

 $s$ is not a leaf

$$1 \;\geq\; 1 \text{ and } H \nless D$$

So, while loop will not execute

$$i \;=\; 2$$

consider node $z$

$$1 \;\neq\; 3$$

B-Tree-Insert-Nonfull $(z, H)$

$$i \;=\; 1$$

 $z$ is a leaf node

$$1 \;\geq\; 1 \text{ and } H \nless G$$

So, while loop will not execute

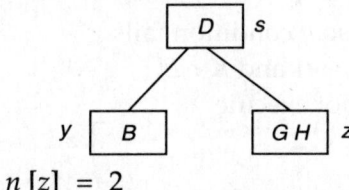

$$n\,[z] \;=\; 2$$

B-Tree-Insert $(T, J)$

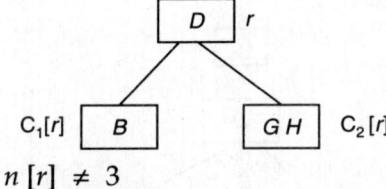

$$n\,[r] \;\neq\; 3$$

B-Tree-Insert-Nonfull $(r, J)$

$$i \;=\; 1$$

 $r$ is not a leaf node

$$1 \;\geq\; 1 \text{ and } J \nless D$$

So, while loop will not execute

$$i \;=\; 2$$

consider $C_2\,[r]$

$$2 \;\neq\; 3$$

B-Tree-Insert-Nonfull $(C_2\,[r], J)$

$$i \;=\; 2$$

$C_2\,[r]$ is a leaf node

$$2 \;\geq\; 1 \text{ and } J \nless H$$

So, while loop will not execute

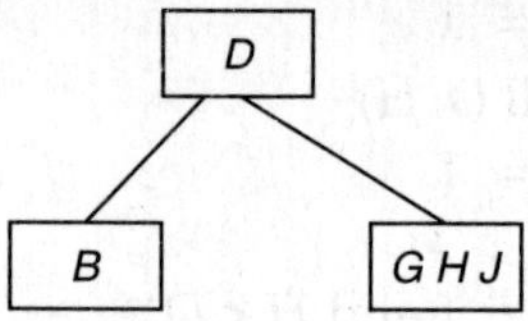

$n\,[C_2\,[r]] = 3$

B-Tree-Insert $(T, K)$

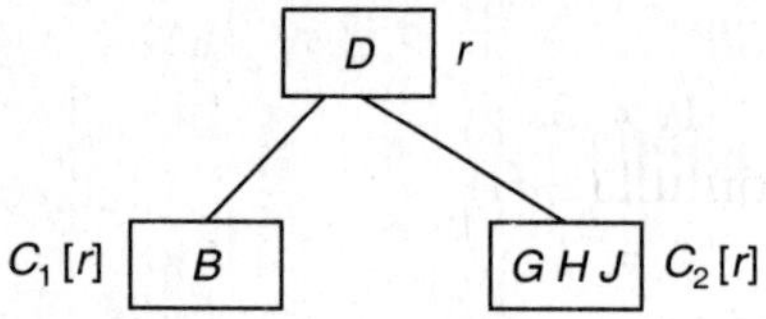

$n\,[r] \neq 3$

B-Tree-Insert-Nonfull $(r, K)$

$i = 1$

$x$ is not a leaf node, so *if* condition fails

$1 \geq 1$ and $K \nless D$

So, while loop will not execute

$i = 2$

Consider $C_2\,[r]$

$n\,[C_2\,[r]] = = 3$

B-Tree-Split-Child $(r, 2)$

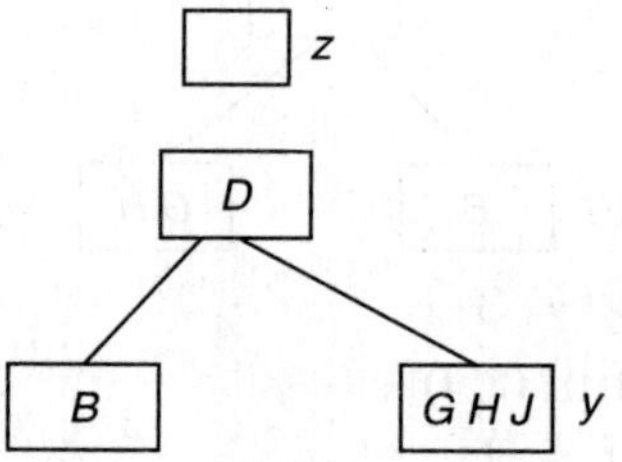

$n\,[z]$ should be 1

$j = 1$

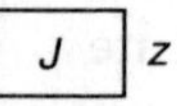

$y$ is a leaf node, so *if* condition fails.

$n\,[y]$ should be 1

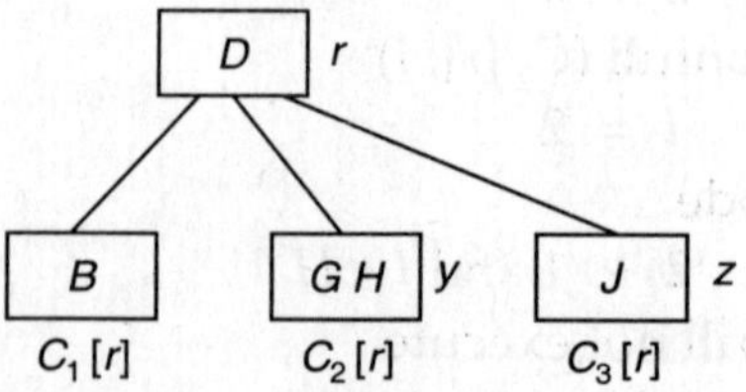

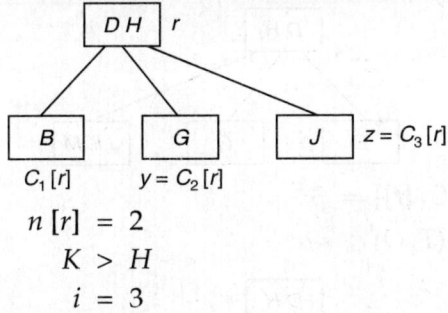

$$n[r] = 2$$
$$K > H$$
$$i = 3$$

B-Tree-Insert-Nonfull ($C_3[r], K$)
$$i = 1$$

$C_3[r]$ is a leaf node
$$1 \geq 1 \text{ and } K \nless J$$

So, while loop will not execute.

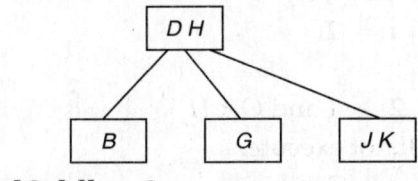

$$n[C_3[r]] = 2$$

B-Tree-Insert ($T, M$)

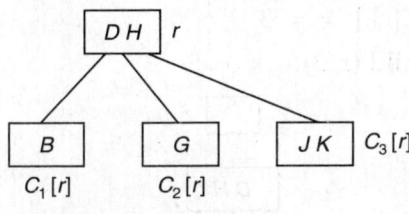

$$n[r] \neq 3$$

B-Tree-Insert-Nonfull ($r, M$)
$$i = 2$$

    $r$ is not a leaf
$$2 \geq 1 \text{ and } M \nless H$$
$$i = 3$$

Disc-Read ($C_3[r]$)
$$n[C_3[r]] \neq 3$$

B-Tree-Insert-Nonfull ($C_3[r], M$)
$$i = 2$$

$C_3[r]$ is a leaf
$$2 \geq 1 \text{ and } M \nless K$$

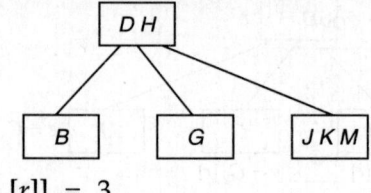

$$n \, [C_3 \, [r]] \; = \; 3$$

B-Tree-Insert $(T, O)$

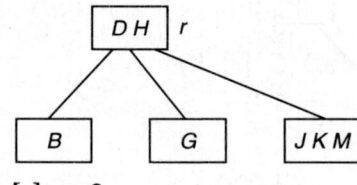

$$n \, [r] \; \neq \; 3$$

B-Tree-Insert-Nonfull $(r, O)$

$$i \; = \; 2$$

$r$ is not a leaf

$$2 \; \geq \; 1 \text{ and } O \nprec H$$

So, while loop will not execute.

$$i \; = \; 3$$

Disc-Read $(C_3 \, [r])$

$$n \, [C_3 \, [r] \, ] \; = \; = \; 3$$

B-Tree-Split-Child $(r, 3)$

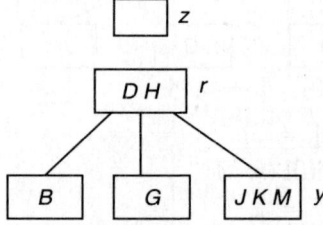

$n \, [z]$ should be 1

$$j \; = \; 1$$

$y$ is a leaf node

$n \, [y]$ should be 1

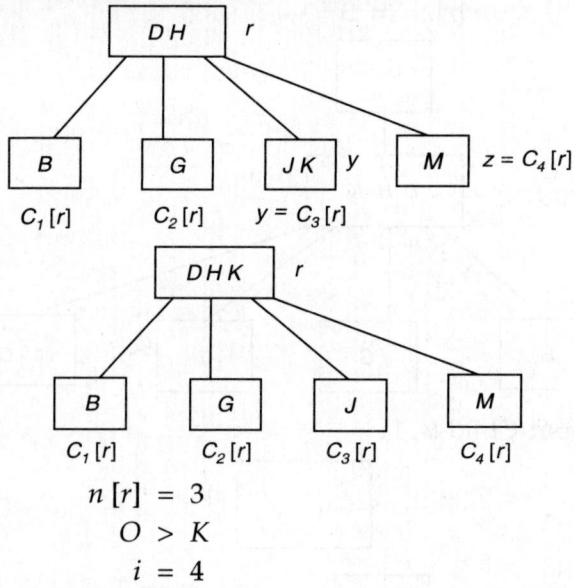

$$n[r] = 3$$
$$O > K$$
$$i = 4$$

B-Tree-Insert-Nonfull $(C_4[r], O)$

$$i = 1$$

$C_4[r]$ is a leaf node

$$1 \geq 1 \text{ and } O \nmid M$$

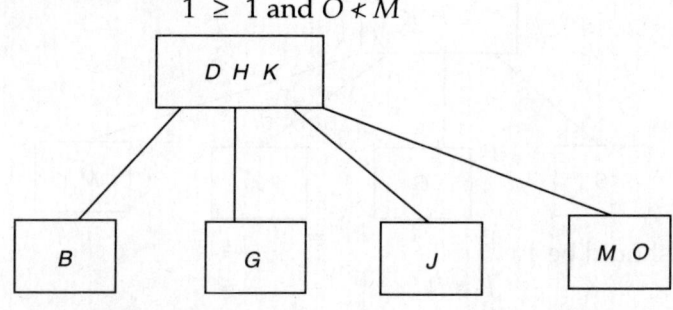

$$n[C_4[r]] = 2$$

Disk-Write $(C_4[r])$

B-Tree-Insert $(T, P)$

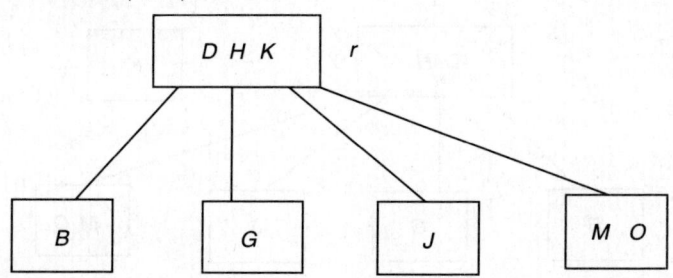

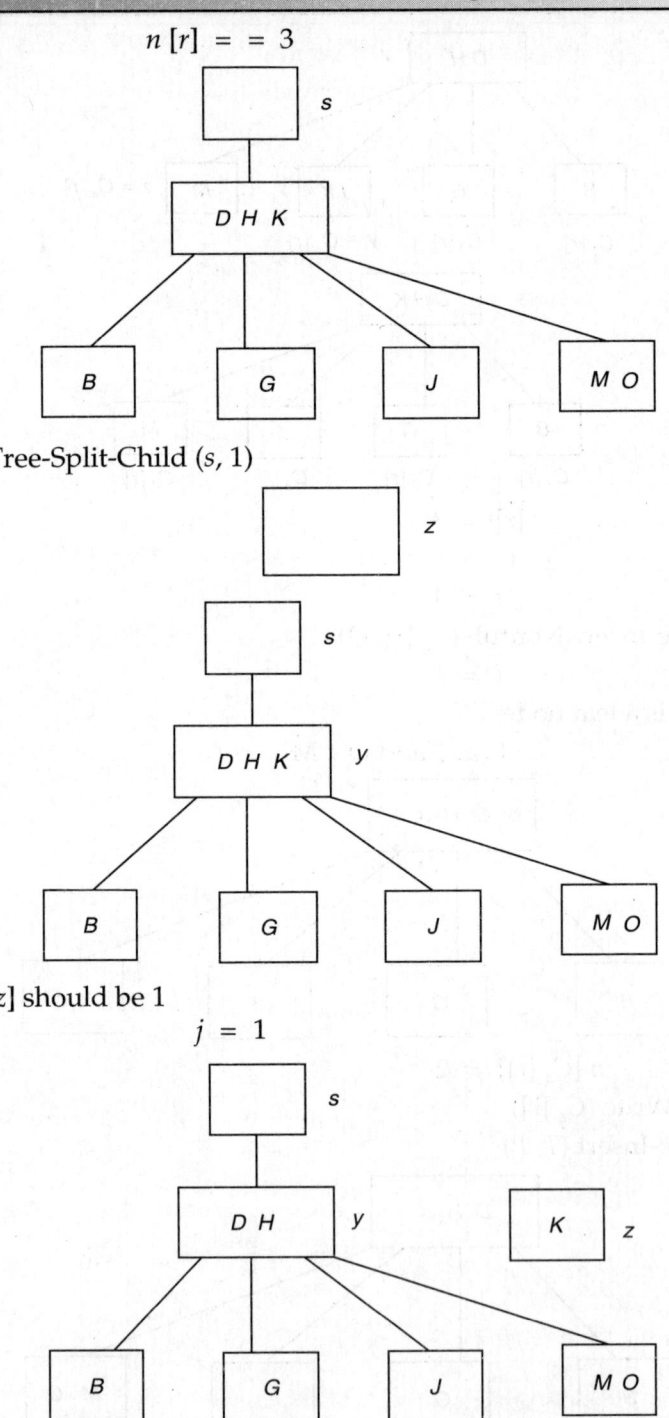

$n[r] == 3$

B-Tree-Split-Child $(s, 1)$

$n[z]$ should be 1

$j = 1$

$y$ is not a leaf node, so *if* condition is true

$j = 1$

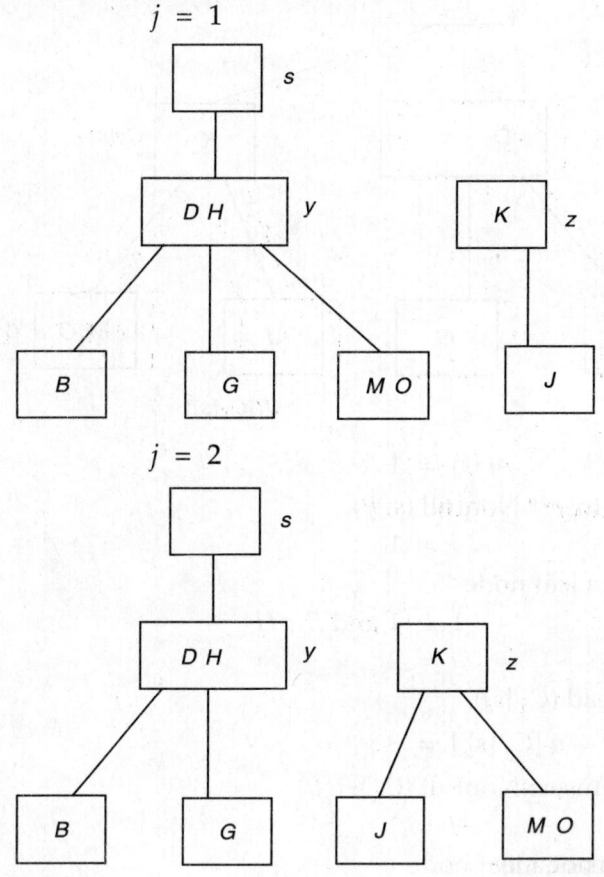

$j = 2$

$n[y]$ should be 1

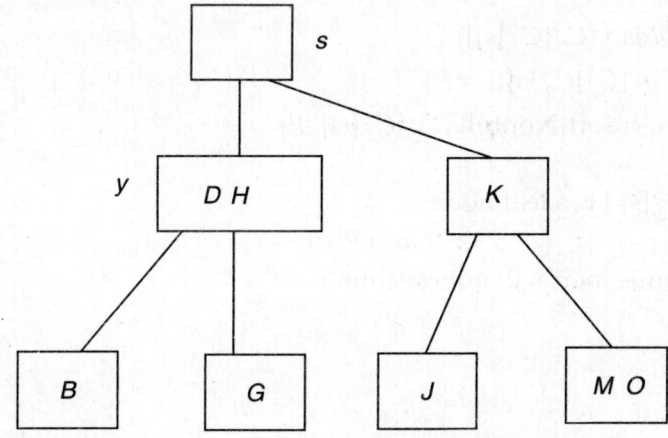

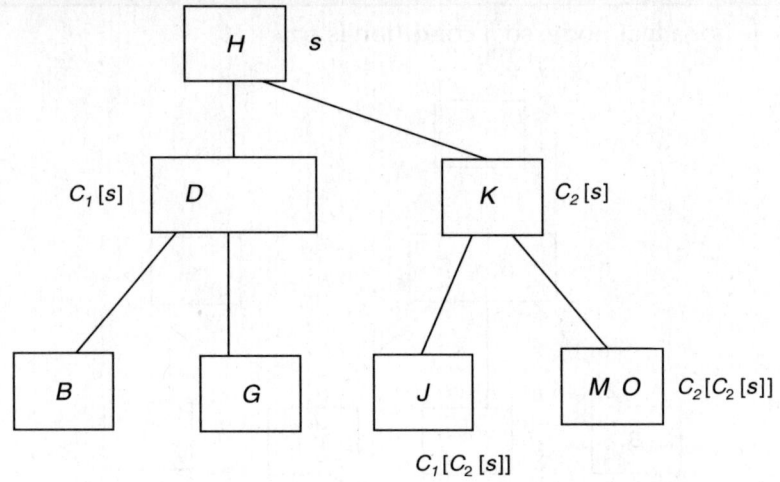

$$n\,[s]\ =\ 1$$

B-Tree-Insert-Nonfull $(s,\,P)$

$$i\ =\ 1$$

$s$ is not a leaf node

$$1\ \geq\ 1 \text{ and } P \nprec H$$
$$i\ =\ 2$$

Disc-Read $(C_2\,[s])$

$$n\,[C_2\,[s]\,]\ \neq\ 3$$

B-Tree-Insert-Nonfull $(C_2\,[s],\,P)$

$$i\ =\ 1$$

$C_2\,[s]$ is not a leaf node

$$1\ \geq\ 1 \text{ and } P \nprec K$$
$$i\ =\ 2$$

Disc-Read $(C_2\,[C_2\,[s]])$

$$n\,[C_2\,[C_2\,[s]]]\ \neq\ 3$$

B-Tree-Insert-Nonfull $(C_2\,[C_2\,[s]],\,P)$

$$i\ =\ 2$$

$C_2\,[C_2\,[s]\,]$ is a leaf node

$$2\ \geq\ 1 \text{ and } P \nprec O$$

So, while loop will not execute

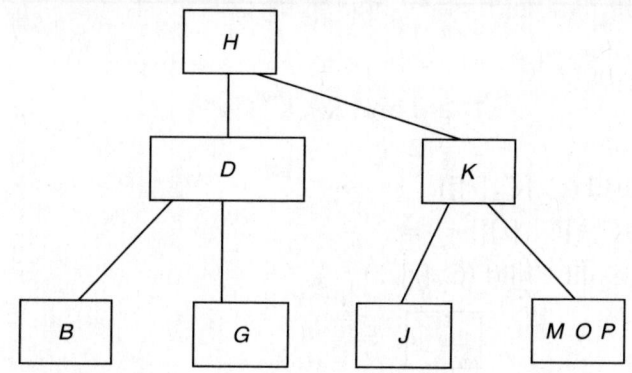

$$n\,[C_2\,[C_2\,[s]]]\ =\ 3$$

Disc-Write $(C_2\,[C_2\,[s]])$

B-Tree-Insert $(T,\,R)$

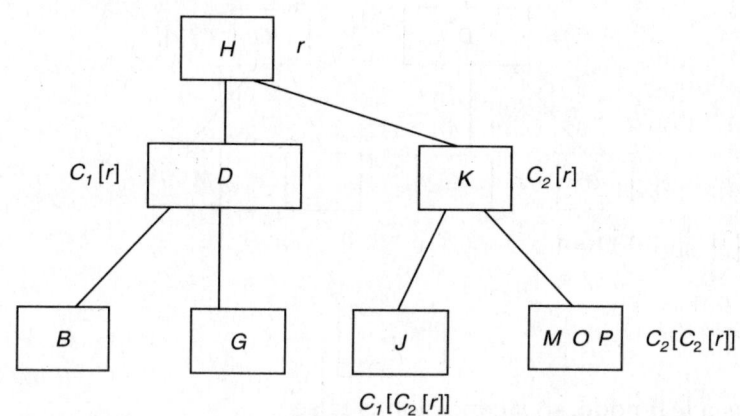

$$n\,[r]\ \neq\ 3$$

B-Tree-Insert-Nonfull $(r,\,R)$

$$i\ =\ 1$$

$r$ is not a leaf node

$$1\ \geq\ 1 \text{ and } R \not< H$$

$$i\ =\ 2$$

Disc-Read $(C_2\,[r])$

$$n\,[C_2\,[r]\,]\ \neq\ 3$$

B-Tree-Insert-Nonfull $(C_2\,[r]),\,R)$

$$i = 1$$

$C_2[r]$ is not a leaf

$$1 \geq 1 \text{ and } R \not< K$$
$$i = 2$$

Disc-Read $(C_2[C_2[r]])$

$n[C_2[C_2[r]]] == 3$

B-Tree-Split-Child $(C_2[r], 2)$

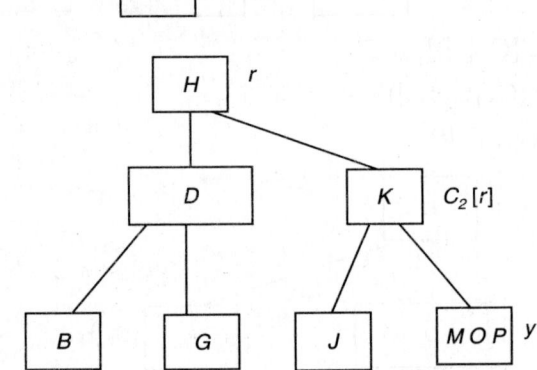

$n[z]$ should be 1

$$j = 1$$

$y$ is a leaf node, so *if* condition is false

$n[y]$ should be 1

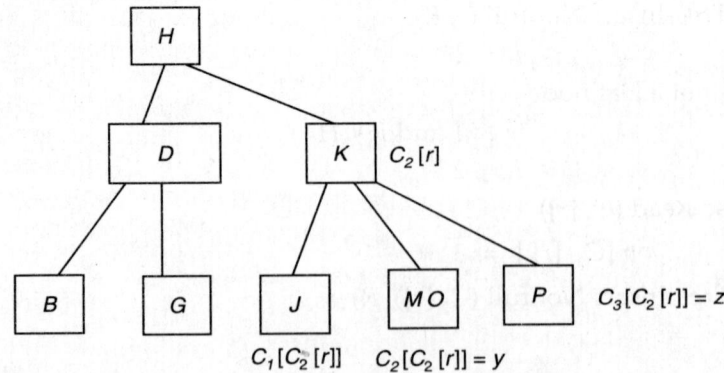

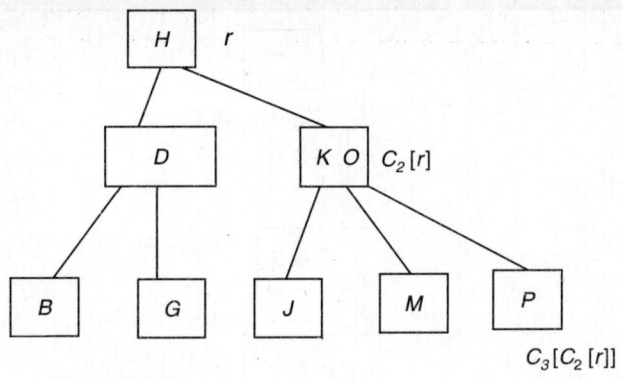

$$n\,[C_2\,[r]] = 2$$
$$R > O$$
$$i = 3$$

B-Tree-Insert-Nonfull $(C_3\,[C_2\,[r]],\,R)$

$$i = 1$$

$C_3\,[C_2\,[r]]$ is a leaf node

$$1 \geq 1 \text{ and } R \not< P$$

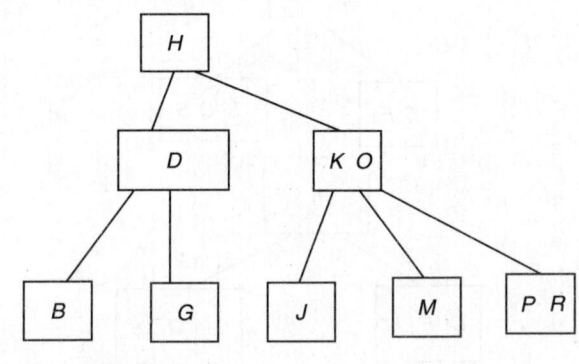

$$n\,[C_3\,[C_2\,[r]]] = 2$$

Disk-write $(C_3\,[C_2\,[r]])$

**Example**

Insert the following keys into an empty B tree with minimum degree 3:

$F, S, Q, K, C, L, H, T, V, W, M, R, N, P, A, B, X, Y, D, Z, E, G, I$

**Solution**

$$t = 3$$
$$2t - 1 = 5$$

Since the minimum degree $t$ for this B-tree is 3, so a node can hold at most 5 keys.

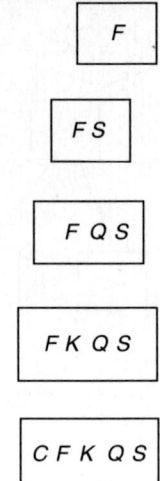

The node *CFKQS* splits into two nodes containing *CF* and *QS*, the key *K* moves up to the root, and *L* is inserted in the *QS* node.

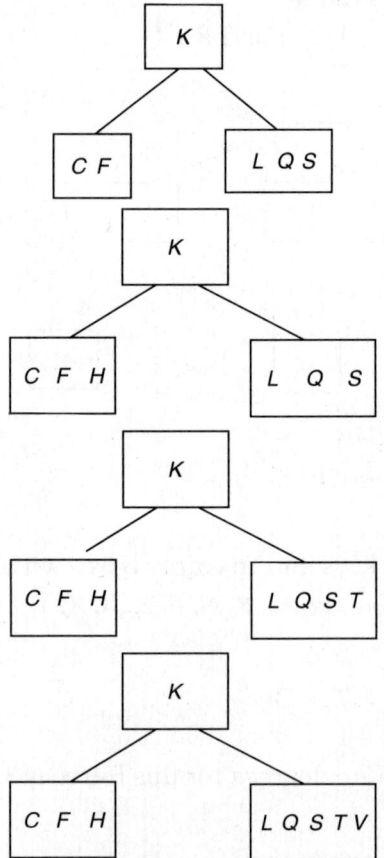

The node *LQSTV* splits into two nodes containing *LQ* and *TV*, the key *S* moves up to the root, and *W* is inserted in the *TV* node.

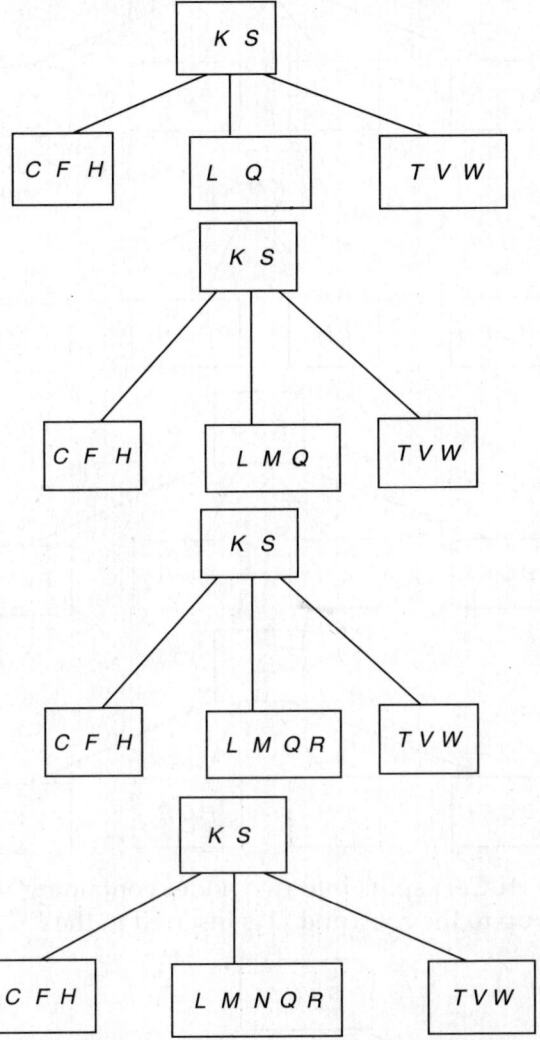

The node *LMNQR* splits into two nodes containing *LM* and *QR*, the key *N* moves up to the root and *P* is inserted in the *QR* node.

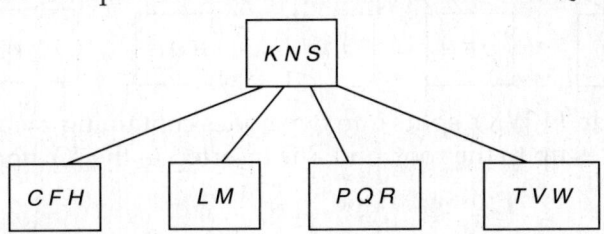

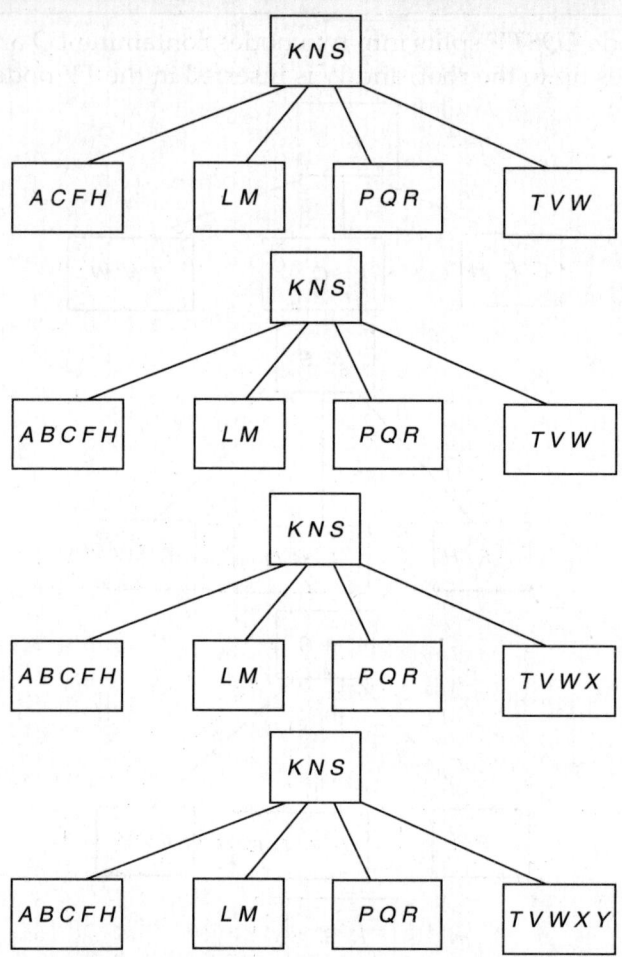

The node *ABCFH* splits into two nodes containing *AB* and *FH*, the key *C* moves up to the root, and *D* is inserted in the *FH* node.

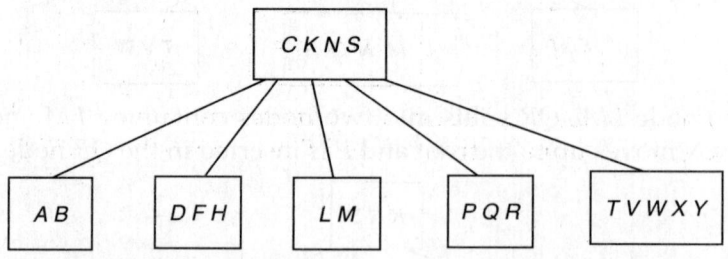

The node *TVWXY* splits into two nodes containing *TV* and *XY*, the key *W* moves up to the root, and *Z* is inserted in the *XY* node.

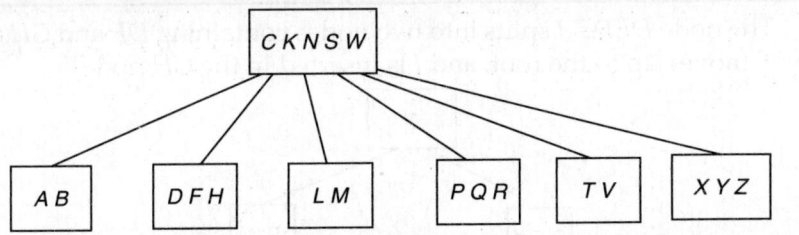

Now, the root splits, since it is full, and the B-tree grows in height by one.

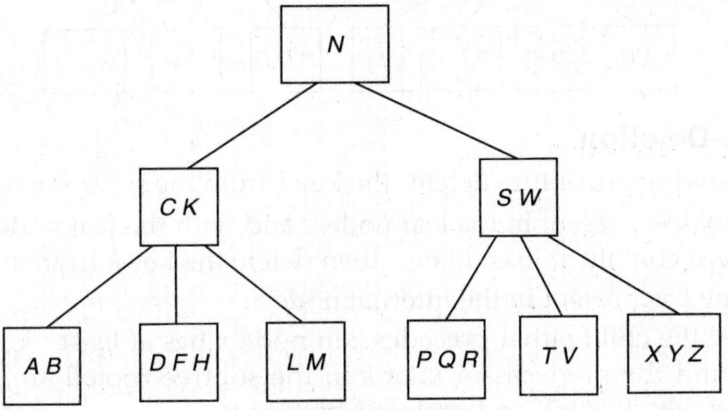

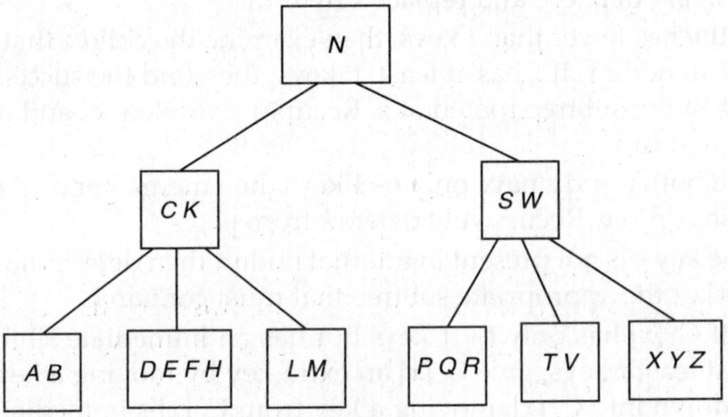

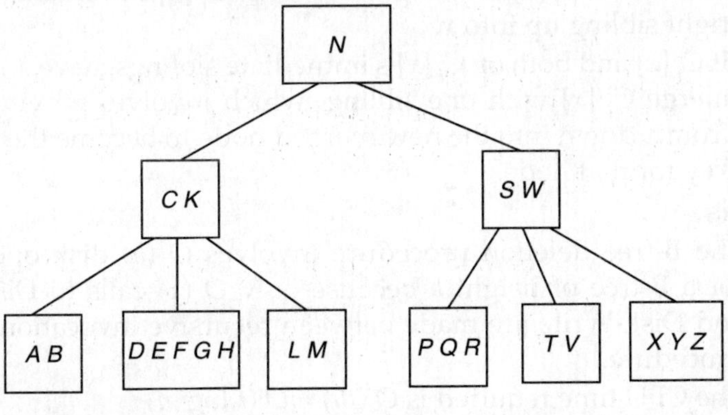

The node *DEFGH* splits into two nodes containing *DE* and *GH*, the key *F* moves up to the root, and *I* is inserted in the *GH* node.

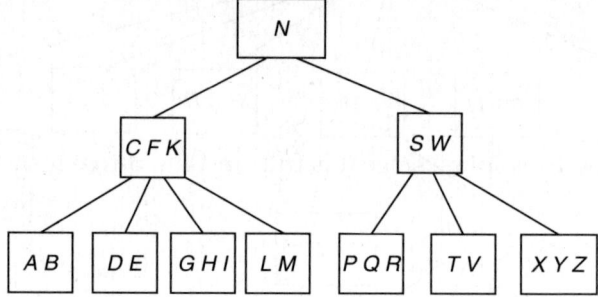

## B-Tree-Deletion

The following procedure deletes the key $k$ from the subtree rooted at $x$:

1. If key $k$ is present in the leaf node $x$ and both the leaf node and its parent contain at least $t$ keys, then delete the key $k$ from $x$.

2. If key $k$ is present in the internal node $x$;

   i.   If the child $y$ that precedes $k$ in node $x$ has at least $t$ keys, then find the predecessor $k'$ of $k$ in the subtree rooted at $y$. Recursively delete $k'$ and replace $k$ by $k'$ in $x$.

   ii.  If $y$ has fewer than $t$ keys, then examine the child $z$ that follows $k$ in node $x$. If $z$ has at least $t$ keys, then find the successor $k'$ of $k$ in the subtree rooted at $z$. Recursively delete $k'$, and replace $k$ by $k'$ in $x$.

   iii. If both $y$ and $z$ have only $t-1$ keys then merge $k$ and all the keys in $z$ into $y$. Recursively delete $k$ from $y$.

3. If the key $k$ is not present in internal node $x$ then determine the root $C_i[x]$ of the appropriate subtree that must contain $k$.

   i.  If $C_i[x]$ has only $t-1$ keys but has an immediate sibling with at least $t$ keys, give $C_i[x]$ an extra key by moving a key from $x$ down into $C_i[x]$, moving a key from $C_i[x]$'s immediate left or right sibling up into $x$.

   ii. If $C_i[x]$ and both of $C_i[x]$'s immediate siblings, have $t-1$ keys, merge $C_i[x]$ with one sibling, which involves moving a key from $x$ down into the new merged node to become the median key for that node.

## Analysis

- The B-tree deletion procedure involves $O(h)$ disk operations for a B-tree of height $h$ because only $O(1)$ calls to Disk-Read and Disk-Write are made between recursive invocations of the procedure.
- The CPU time required is $O(th) = O(t \log_t n)$

## Example

Delete the keys $F$, $M$, $G$, $D$, $B$, $C$, $P$, $V$ from the following B-tree with $t = 3$.

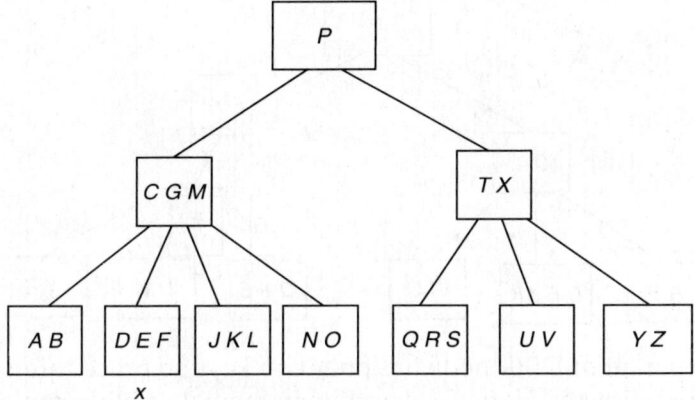

## Solution

Apply procedure (1) for deleting key $F$

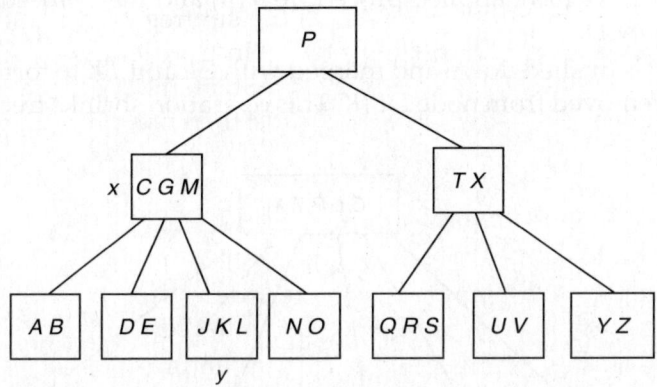

Child $y$ has three keys. The predecessor $L$ of $M$ moves up to take $M$'s position. So, we have applied procedure 2(i) for deleting key $M$.

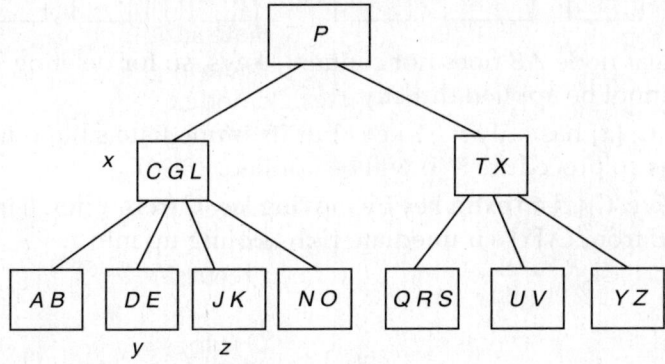

Child $y$ and child $z$ both have $t - 1$ keys, so apply procedure 2 (iii) and then procedure 1. Merge $G$ and all the keys in $z$ into $y$ and then delete key $G$ from $y$.

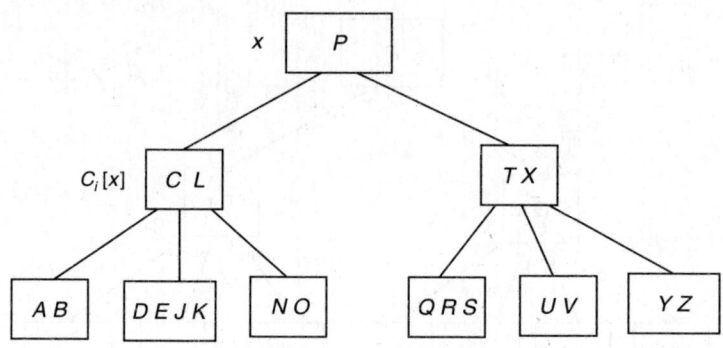

The parent of $DEJK$ node has only two keys so procedure 1 cannot be applied directly. Here $C_i[x]$ and its immediate sibling contain $t - 1$ keys so merge $C_i[x]$ with the sibling which involves moving a key from $x$ down into the new merged node to become the median key for that node. Thus, we have applied procedure 3 (ii) and then procedure 1 for deleting key $D$.

Key $P$ is pushed down and merged with $CL$ and $TX$ to form $CLPTX$ and $D$ is removed from node $DEJK$. This operation shrinks tree's height by 1.

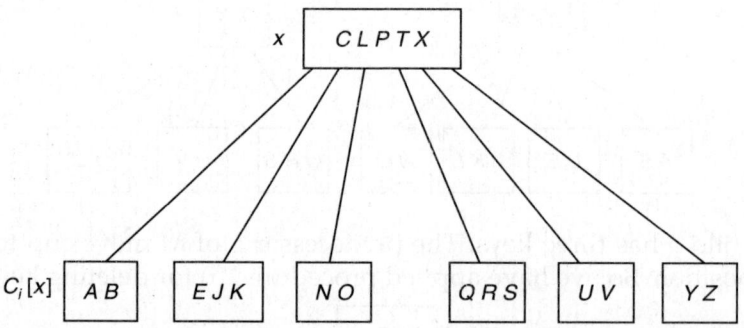

The leaf node $AB$ does not contain $t$ keys, so for deleting $B$ procedure 1 cannot be applied directly.

Here $C_i[x]$ has only $t - 1$ keys but its immediate sibling node $EJK$ has 3 keys so procedure 3 (i) will be applied.

We give $C_i[x]$ an extra key by moving key $C$ from $x$ down into $C_i[x]$ and key $E$ from $C_i[x]$'s immediate right sibling up into $x$.

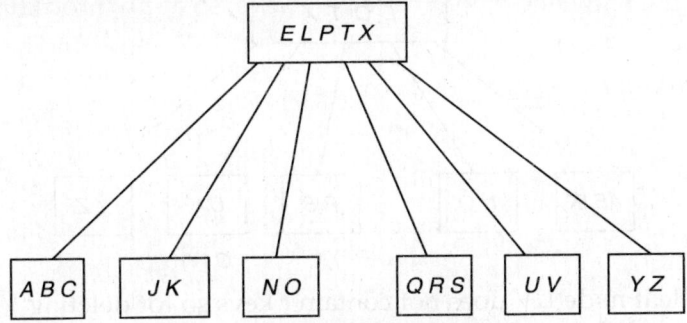

Then apply procedure 1 for deleting key B.

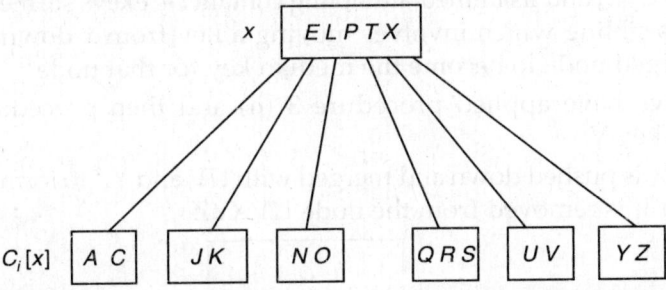

The leaf node AC does not contain $t$ keys so, for deleting C procedure 1 cannot be applied directly.

Here $C_i[x]$ and its immediate sibling contain $t-1$ keys so merge $C_i[x]$ with this sibling which involves moving a key from $x$ down into the new merged node to become the median key for that node.

Thus, we have applied procedure 3 (ii) and then procedure 1 for deleting key C.

Key E is pushed down and merged with AC and JK to form ACEJK and then C is removed from the node ACEJK.

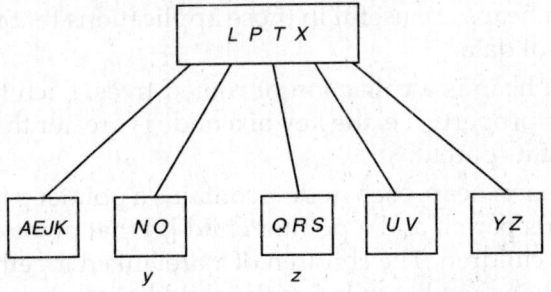

child $y$ has two keys and child $z$ has three keys. Key Q is the successor of key P.

The successor Q of P moves up to take P's position.

So, we have applied procedure 2 (ii) for deleting key P.

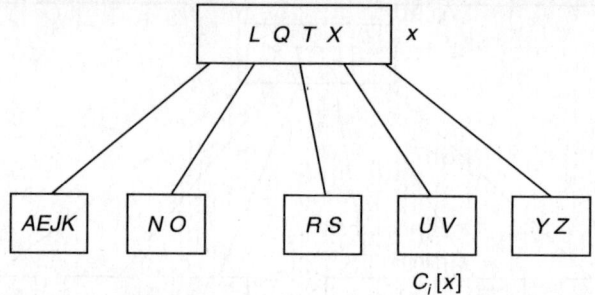

$$C_i[x]$$

The leaf node $UV$ does not contain $t$ keys so for deleting $V$, procedure 1 cannot be applied directly.

Here $C_i[x]$ and its immediate sibling contain $t-1$ keys so, merge $C_i[x]$ with this sibling which involves moving a key from $x$ down into the new merged node to become the median key for that node.

So, we have applied procedure 3 (ii) and then procedure 1 for deleting key V.

Key $X$ is pushed down and merged with $UV$ and $YZ$ to form $UVXYZ$ and then $V$ is removed from the node $UVXYZ$.

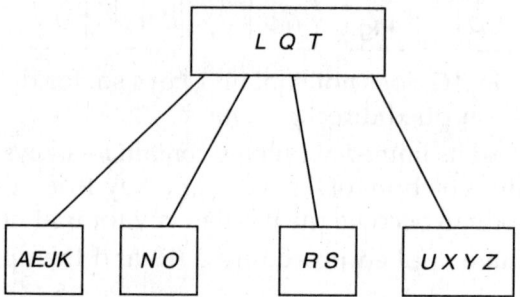

## 2.7 FIBONACCI HEAPS

- Fibonacci heaps are useful in those applications that manage large amounts of data.
- Fibonacci heap is a collection of rooted trees. Each tree follows a min-heap property, i.e. the key of a node is greater than or equal to the key of its parent.
- In a Fibonacci heap, each node $x$ contains a pointer $p[x]$ that points towards its parent and a pointer child $[x]$ that points towards any one of its children. The children of $x$ are linked together in a circular, doubly linked list called as the child list of $x$.
- Each child $y$ in the child list has pointers $y[left]$ and $y[right]$ that points to $y$'s, left and right siblings respectively.
- Siblings appear in a child list in any order.

- Each node $x$ has two attributes, degree $[x]$ and mark $[x]$.
- The roots of all the trees in a Fibonacci heap are linked together using their left and right pointers into a circular, doubly linked list called the root list of the Fibonacci heap.
- Pointer min [H] points to the node in the root list whose key is minimum.

Figure 2.1 shows a Fibonacci heap.

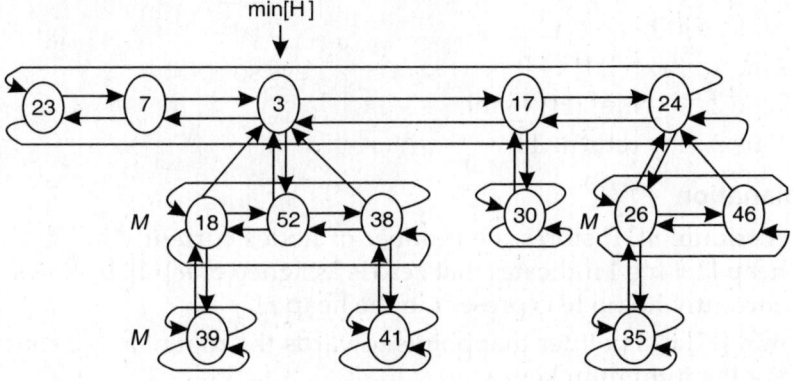

**Fig. 2.1:** Fibonacci heap

## Potential Function:

Potential method is used to analyze the performance of Fibonacci heap operations.

Consider a Fibonacci heap H

Let     $t$ (H) be the number of trees in H and

        $m$ (H) be the number of marked nodes in H

then the potential function $\varphi$ (H) of Fibonacci heap H is

$$\varphi (H) = t (H) + 2m (H)$$

## Example

Determine the potential function of the given figure:

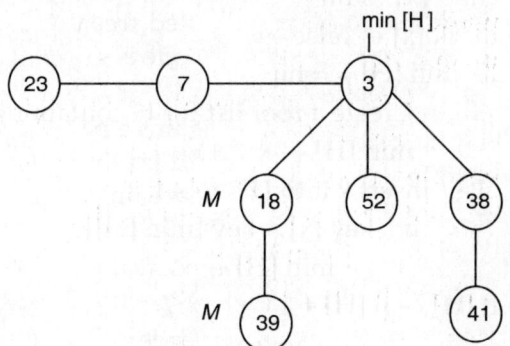

**Solution**

$$\varphi(H) = t(H) + 2\,m(H)$$
$$= 3 + (2 \times 2)$$
$$= 7$$

**Algorithm**

Creating new Fibonacci Heap

    Make-Fib-heap ( )

    Line 1       $n[H] \leftarrow 0$

    Line 2       $min[H] \leftarrow nil$

    Line 3       return  H

**Explanation**

- Attribute $n[H]$ stores the number of nodes currently in Fibonacci heap H. Line 1 indicates that zero is assigned to $n[H]$. It shows that currently no node is present in the heap H.

- $min[H]$ is a pointer that points towards the root of a tree containing the minimum key.

  Line 2 indicates that $min[H]$ is nil which means that there are no trees in H.

- Line 3 returns the empty Fibonacci heap H.

**Analysis**

- Since $t(H) = 0$ and $m(H) = 0$, therefore $\varphi(H) = 0$.

**Algorithm**

Inserting a node into a Fibonacci heap H

Fib-heap-insert (H, x)

    Line 1       degree $[x] \leftarrow 0$

    Line 2       $p[x] \leftarrow nil$

    Line 3       child $[x] \leftarrow nil$

    Line 4       mark $[x] \leftarrow false$

    Line 5       if  $min[H] = = nil$

    Line 6                Create a root list for H containing just x

    Line 7                $min[H] \leftarrow x$

    Line 8       else   insert x into H's root list

    Line 9            if   key $[x] <$ key $[min[H]]$

    Line 10                $min[H] \leftarrow x$

    Line 11      $n[H] \leftarrow n[H] + 1$

## Explanation

This algorithm assumes that the node has already been allocated and key [x] has already been filled in.

- $x$ is a node. key [x] denotes the key of node $x$.

  This node $x$ is going to be inserted in the Fibonacci heap H.

- The attribute degree [x] stores the number of children of node $x$. Line 1 indicates that 0 is assigned as the degree of $x$. It shows that node $x$ has no children.

- $p$ [x] is a pointer that points towards the parent of node $x$. Line 2 indicates that $p$ [x] is nil. This means that $x$ will be a root node.

- child [x] is a pointer that points towards any of the children of node $x$. Line 3 indicates that child [x] is nil which means that the pointer is pointed towards nothing because node $x$ has no children.

- Line 4 indicates that a false value is assigned to mark [x]. It means that node $x$ remains unmarked.

  The Boolean valued attribute mark [x] indicates whether or not node $x$ has lost a child since the last time node $x$ was made the child of another node.

- Line 5 indicates that the *if* condition is checked. If this condition is true, i.e. there are no trees in H then the execution of Line 6 and Line 7 takes place. If this condition is false, the algorithm skips the execution of Line 6 and Line 7 and goes directly for the execution of Line 8.

- Line 6 creates a root list for Fibonacci heap H that contains just a single root node $x$.

- Line 7 indicates that the pointer min [H] is made to point towards the node $x$.

- Line 8 indicates the insertion of node $x$ into H's already present root list.

- Line 9 indicates that if the key value of node $x$ is smaller than the key value of that node at which the pointer min [H] points to, then the execution of Line 10 takes place.

- Line 10 indicates that the pointer min [H] is made to point to node $x$.

- Line 11 indicates that the value of $n$ [H] is incremented by 1. It is done due to the insertion of new node $x$ into heap H.

## Analysis

- Let H be the input Fibonacci heap and H' be the resulting Fibonacci heap.

$$t (H') = t (H) + 1$$
$$m (H') = m (H)$$

The increase in potential will be

$$((t (H) +1) + 2 m (H)) - (t (H) + 2 m (H)) = 1$$

Since the actual cost is $O (1)$, the amortized cost will be $O (1) + 1$ $= O (1)$.

## Example

Insert a node with key 10 into the following Fibonacci heap.

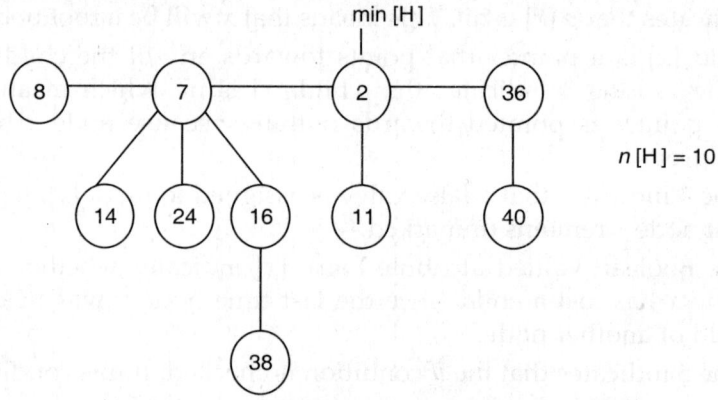

## Solution

$$min [H] \neq nil$$

Here min [H] is pointing towards the node with key 2

So, the execution of Line 8 takes place

$x$ is inserted into $H'$s root list.

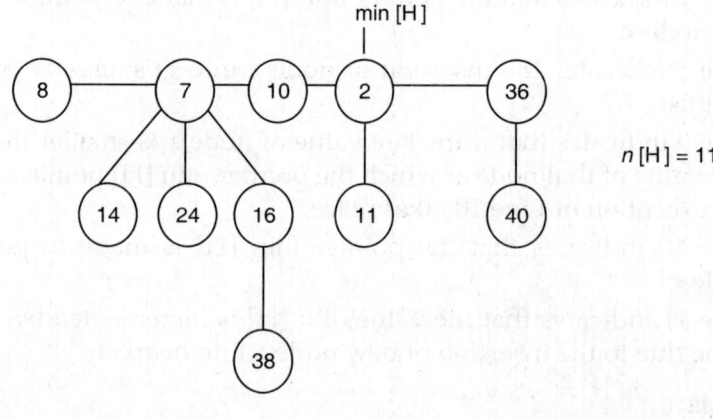

$$10 \nless 2$$

So, *if* condition fails

## Algorithm

Finding the minimum node.

The minimum node of a Fibonacci heap H is given by the pointer min [H].

## Analysis

- The minimum node can be found in $O(1)$ actual time.
- Since the potential of H does not change, therefore the amortized cost of the above procedure is equal to its $O(1)$ actual cost.

## Algorithm

Uniting two Fibonacci heaps

Fib-Heap-Union ($H_1$, $H_2$)

| | |
|---|---|
| Line 1 | H ← Make-Fib-heap ( ) |
| Line 2 | min [H] ← min [$H_1$] |
| Line 3 | concatenate the root lists of $H_1$ and $H_2$ with the root list of H. |
| Line 4 | if (min [$H_1$] = = nil) or (min [$H_2$] ≠ nil and key [min [$H_2$]] < key[min [$H_1$]]) |
| Line 5 | min [H] ← min [$H_2$] |
| Line 6 | n [H] ← n [$H_1$] + n [$H_2$] |
| Line 7 | return H |

## Explanation

- Line 1 indicates that a Fibonacci heap H is created by executing Make-Fib-heap ( ) algorithm.
- Line 2 indicates that the pointer min [H] is made to point to that node which was earlier pointed by pointer min [$H_1$].
- Line 3 concatenates the root lists of $H_1$ and $H_2$ with the root list of H.
- Line 4 indicates that *if* condition is true when either one or both cases are true.

    Case i.    There are no trees in $H_1$, i.e. Fibonacci heap $H_1$ is empty.

    Case ii.    Fibonacci heap $H_2$ is not empty and the key value for the node pointed by min [$H_2$] pointer is less than the key value for the node pointed by min [$H_1$] pointer.

    If the *if* condition is true then the execution of Line 5 takes place.

- Line 5 indicates that the pointer min [H] is made to point to that node which was earlier pointed by pointer min [$H_2$].
- Line 6 indicates that the number of nodes in the Fibonacci heap H is the sum of the number of nodes in Fibonacci heaps $H_1$ and $H_2$.
- Line 7 returns the resulting Fibonacci heap H.

## Analysis

- Here $t(H) = t(H_1) + t(H_2)$ and

  $m(H) = m(H_1) + m(H_2)$

  Therefore,

  the change in potential will be

  $\varphi(H) - (\varphi(H_1) + \varphi(H_2))$

  $= (t(H) + 2\, m(H)) - ((t(H_1) + 2\, m(H_1)) + (t(H_2)$

  $+ 2\, m(H_2)))$

  $= (t(H_1) + t(H_2) + 2\, m(H_1) + 2\, m(H_2)) - ((t(H_1)$

  $+ 2\, m(H_1)) + (t(H_2) + 2\, m(H_2)))$

  $= 0$

- Therefore, the amortized cost of Fib-Heap-Union is equal to its $O(1)$ actual cost.

## Example

Apply union operation on the given Fibonacci heaps.

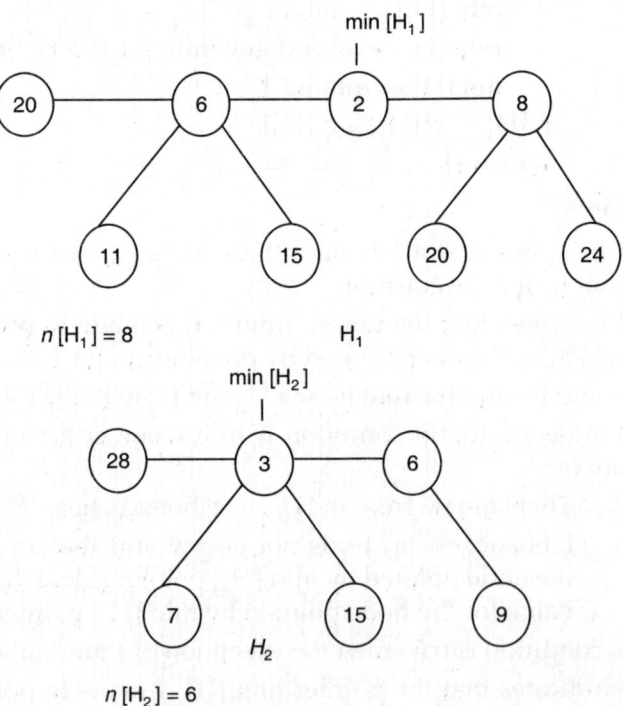

$n[H_1] = 8$                                    $H_1$

$n[H_2] = 6$

## Solution

$$\min[H] = \min[H_1]$$

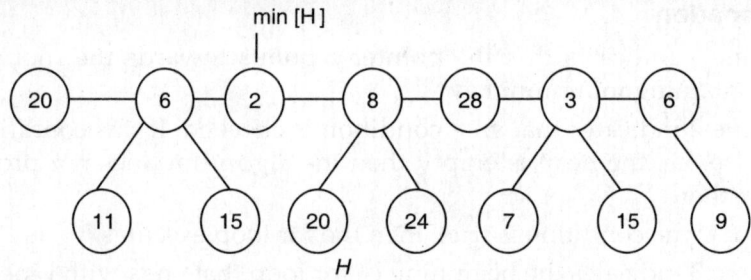

Here

Case i.        min $[H_1] \neq$ nil

Case ii.       min $[H_2] \neq$ nil and 3 ∢ 2

Both cases are false

Since none of them is true, therefore, the *if* condition fails

Hence, there is no execution of Line 5

$$n [H] = 8 + 6 = 14$$

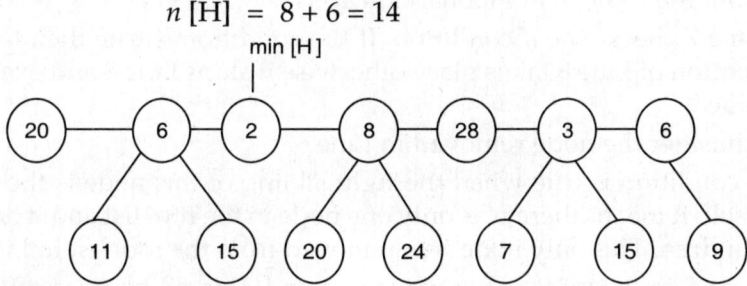

## Algorithm

Extracting the minimum key

    Fib-heap-Extract-Min (H)

    Line 1      z ← min [H]
    Line 2      if  z ≠ nil
    Line 3            for each child x of z
    Line 4                  add x to the root list of H
    Line 5                  p [x] ← nil
    Line 6            remove z from the root list of H
    Line 7            if    z = = right [z]
    Line 8                  min [H] ← nil
    Line 9            else  min [H] ← right [z]
    Line 10                 Consolidate (H)
    Line 11           n [H] ← n [H] – 1
    Line 12     return z

## Explanation

- Line 1 indicates that the pointer $z$ points towards the root node containing minimum key.
- Line 2 indicates that an *if* condition is checked. If this condition is false, i.e. the heap is empty then the algorithm does not proceed further.

  But if the condition is true, then the for loop executes.
- Line 3 indicates the beginning of for loop that ends with Line 5.

  The for loop is applicable for each child $x$ of the node pointed by pointer $z$.
- Line 4 indicates that child $x$ is added to the root list of H.
- Line 5 indicates that $p[x]$ is nil which means that the child node $x$ is now made the root node.
- Line 6 indicates that the root node pointed by pointer $z$ is removed from the root list of Fibonacci heap H.
- Line 7 checks the *if* condition. If the condition is true then the execution of Line 8 takes place otherwise it skips Line 8 and executes Line 9.

  Consider the node removed in Line 6.

  *if* condition is true when the right sibling of this node is the node itself. It means there was only one node in the root list and it had no children. This only node was removed from the root list in Line 6.
- Line 8 indicates that the value of min [H] is nil because after the removal of the only node in Line 6, the heap becomes empty.
- Line 9 indicates that the pointer min [H] now points towards the right sibling of that node which was removed in Line 6. This right sibling may not contain the minimum key.
- Line 10 calls the procedure Consolidate (H).
- Line 11 indicates that the total number of nodes in the Fibonacci heap H is decreased by 1.
- Line 12 returns the pointer $z$ which points towards the root node containing the minimum key.

## Algorithm

```
Consolidate (H)
Line 1     let A [0... D (n [H])] be a new array
Line 2     for  i ← 0 to D (n [H])
Line 3          A [i] ← nil
Line 4     for each node w in the root list of H
Line 5          x ← w
```

Line 6            $d \leftarrow$ degree [x]

Line 7          while A [d] $\neq$ nil

Line 8               $y \leftarrow$ A [d]

Line 9             if key [x] > key [y]

Line 10                 exchange x with y

Line 11             Fib-heap-link (H, y, x)

Line 12             A [d] $\leftarrow$ nil

Line 13             $d \leftarrow d + 1$

Line 14       A [d] $\leftarrow$ x

Line 15      min [H] $\leftarrow$ nil

Line 16      for i $\leftarrow$ 0 to D (n [H])

Line 17        if A [i] $\neq$ nil

Line 18           if min [H] = = nil

Line 19               create a root list for H containing just A [i]

Line 20               min [H] $\leftarrow$ A [i]

Line 21         else        insert A [i] into H's root list

Line 22              if key [A[i]] < key [min [H]]

Line 23               min [H] $\leftarrow$ A [i]

## Explanation

- Line 1 indicates that an array A has been taken. It has the upper limit equal to the maximum degree $D(n)$ of any node in an $n$-node Fibonacci heap H. $D(n) \leq \lfloor \lg n \rfloor$, where $n$ is the total number of nodes in the Fibonacci heap H. This array A is used to keep track of roots according to their degrees.

- Line 2 indicates the beginning of for loop that ends with Line 3. This loop is applicable for each element of array A.

- Line 3 indicates that each entry of array A is initialized by assigning nil to them.

- Line 4 indicates the beginning of for loop that ends with Line 14. This loop is applicable for each node $w$ in the root list of H. The for loop processes the root list by starting with the node pointed by min [H] and following right pointers.

- Line 5 indicates node $w$ and node $x$ are same. Node $w$ keeps on changing after every iteration of for loop of Line 4.

- Line 6 indicates that variable $d$ stores the degree of node $x$.

- Line 7 indicates the beginning of while loop that ends with Line 13. This loop terminates when A[$d$] = = nil, i.e. there is no other root with the same degree as node $x$.

In general, if $A[i] = y$, then $y$ is currently a root with degree $[y] = i$.

- Line 8 indicates that $y$ is currently a root with degree $[y] = d$. So $y$ is the node with the same degree as $x$.
- Line 9 checks the *if* condition. This condition is true when key $[x]$ is greater than key $[y]$. If this condition is true then the execution of Line 10 takes place otherwise it skips.
- Line 10 exchanges node $x$ with node $y$.
- Line 11 calls the procedure Fib-heap-link (H, $y$, $x$).
- Line 12 indicates that since node $y$ is no longer a root so the pointer to it in array A has been removed. Hence $A[d]$ is nil.
- Line 13 indicates that the degree of $x$ is incremented by 1.
- Line 14 indicates that A $[d]$ is set to point to node $x$.
- Line 15 indicates that the root list becomes empty.
- Line 16 indicates the beginning of for loop that ends with Line 23. This loop is applicable for each element of array A.
- Line 17 checks the *if* condition. This condition is true until the array element is non-empty. Failing of this condition leads to the execution of next iteration of for loop of Line 16.
- Line 18 checks the *if* condition. This condition is true if min [H] is nil, i.e. the Fibonacci heap is empty.
- Line 19 creates a root list for H containing just A $[i]$.
- Line 20 indicates that the pointer min [H] is made to point to that node which is pointed by A $[i]$.
- When the *if* condition of Line 18 is false, Line 21 inserts the node pointed by A $[i]$ into H's root list.
- Line 22 checks the *if* condition. This condition is true if key of A $[i]$ is less than the key of min [H].
- Line 23 indicates that the pointer min [H] is made to point to that node which is pointed by A $[i]$.

## Algorithm

Fib-heap-link (H, y, x)

    Line 1       remove y from the root list of H.

    Line 2       make y a child of x, incrementing degree [x]

    Line 3       mark [y] ← False, i.e. clearing the mark on y.

## Analysis

- The for loop of Fib-heap-Extract-Min procedure and Lines 2-3 and 16-23 of Consolidate procedure contribute $O(D(n))$ to the actual cost of extracting the minimum node.

The for loop of Lines 4-14 of Consolidate procedure contributes $O(D(n) + t(H))$ to the actual cost of extracting the minimum node. Within a given iteration of the for loop of Lines 4-14, the number of iterations of the while loop of Lines 7-13 depends on the root list. But, every time through the while loop, one of the roots is linked to another. Thus, the total number of iterations of the while loop over all iterations of the for loop is at most the number of roots in the root list. Hence the total amount of work performed in the for loop is at most proportional to $D(n) + t(H)$.

- So, the total actual work in extracting the minimum node is $O(D(n) + t(H))$.
- The potential before extracting the minimum node is $t(H) + 2m(H)$, and the potential afterward is at most $(D(n) +1) + 2m(H)$ because at most $D(n) + 1$ roots remain and no nodes become marked during the operation.
- Thus, the amortized cost is at most

$$O(D(n) + t(H)) + ((D(n) +1) + 2m(H)) - (t(H) + 2m(H))$$
$$= O(D(n)) + O(t(H)) - t(H)$$
$$= O(D(n)),$$

because we can scale up the units of potential to dominate the constant hidden in $O(t(H))$.

- The cost of performing each link is paid for by the reduction in potential due to the link's reducing the number of roots by one.

## Example

Apply Fib-heap-extract-min operation on the following Fibonacci heap. $M$ denotes marked nodes.

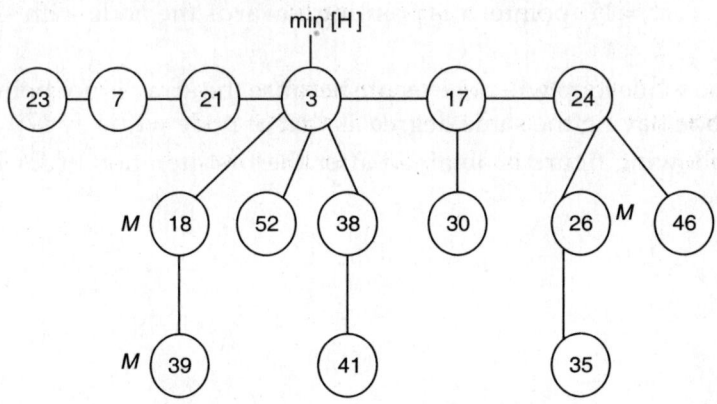

## Solution

$$n[H] = 15$$
$$z = 3$$

Following figure is obtained after the execution of Line 9

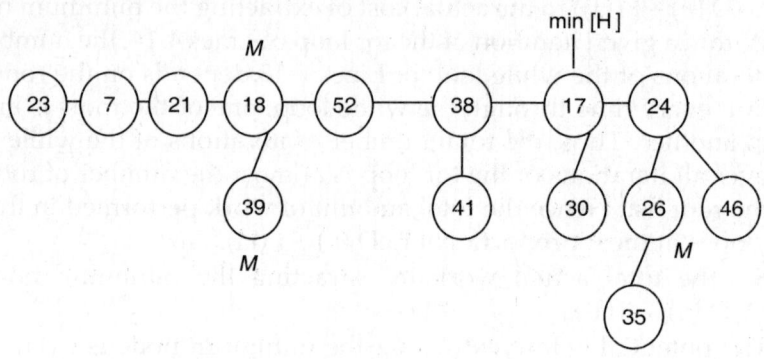

$$D\,[n] = \lfloor \lg n \rfloor$$
$$= \lfloor \lg 15 \rfloor$$
$$= \lfloor 3.9 \rfloor$$
$$= 3$$

```
     0   1   2   3
   ┌───┬───┬───┬───┐
   │ / │ / │ / │ / │
   └───┴───┴───┴───┘
         Array
```

The for loop of Consolidate (H) procedure beginning from Line 4 is applicable for the following values of $w$.

$$w = 17, 24, 23, 7, 21, 18, 52, 38$$

In this order the values of $w$ changes after each iteration of for loop.

$w = 17$ (pointer $w$ is pointing towards the node with key 17)

$x = 17$ (pointer $x$ is pointing towards the node with key 17)

$d = 1$

The while loop will not execute because the array is not pointing to any node having the same degree as that of node with key 17.

Following figure is obtained after the first iteration of for loop of Line 4.

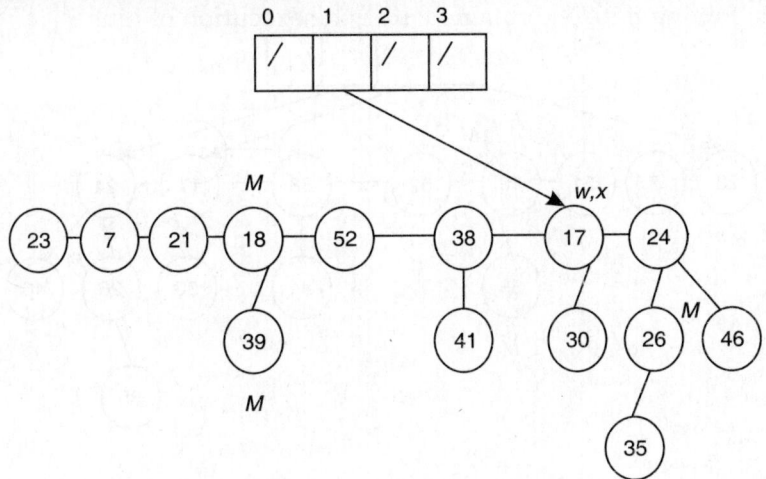

$$w = 24$$
$$x = 24$$
$$d = 2$$

while loop will not execute

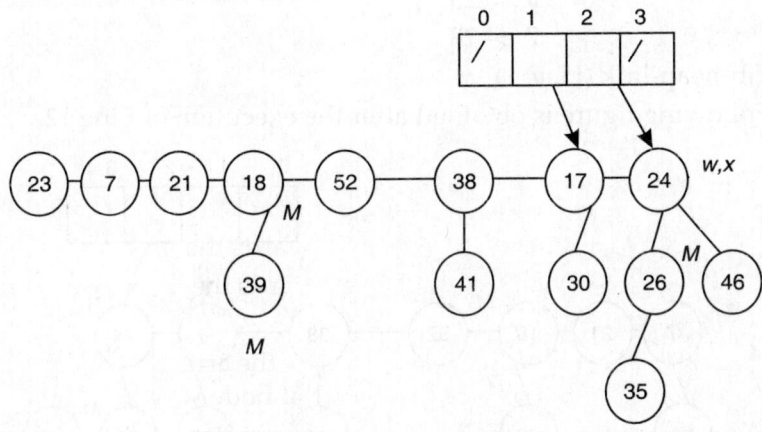

$$w = 23$$
$$x = 23$$
$$d = 0$$

while loop will not execute

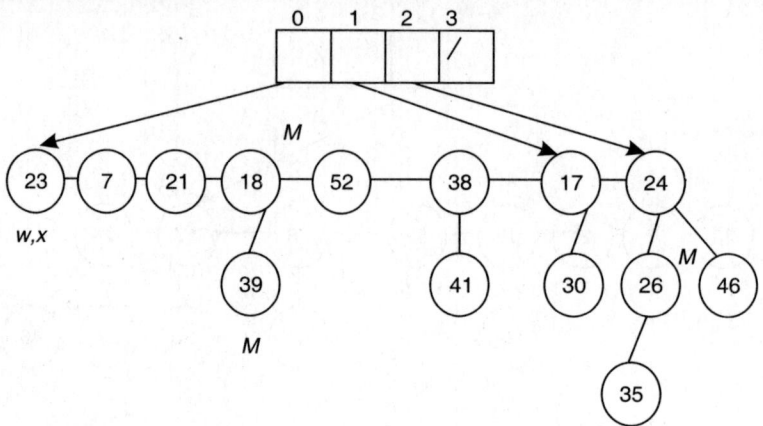

$$w = 7$$
$$x = 7$$
$$d = 0$$

while loop executes because the array is pointing towards a node having same degree as that of node $x$.

$$y = 23$$
$$7 \not> 23$$

Fib-heap-link (H, $y$, $x$)

Following figure is obtained after the execution of Line 12.

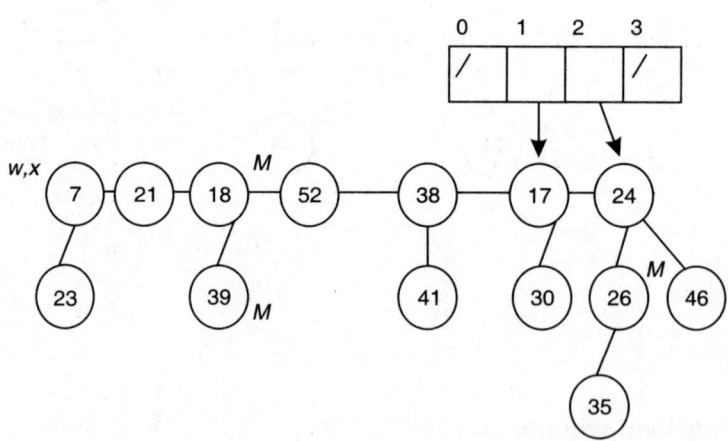

$$d = 1$$

For second iteration of while loop

$$y = 17$$

After the execution of Line 12

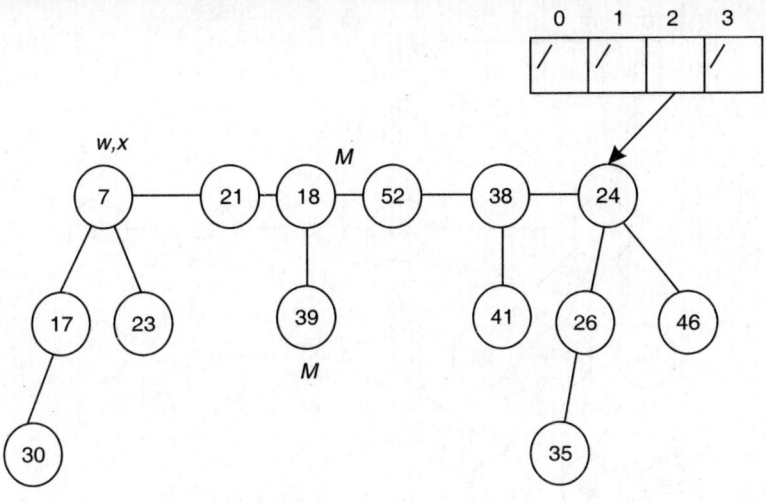

$$d = 2$$

For third iteration of while loop

$$y = 24$$

After the execution of Line 12

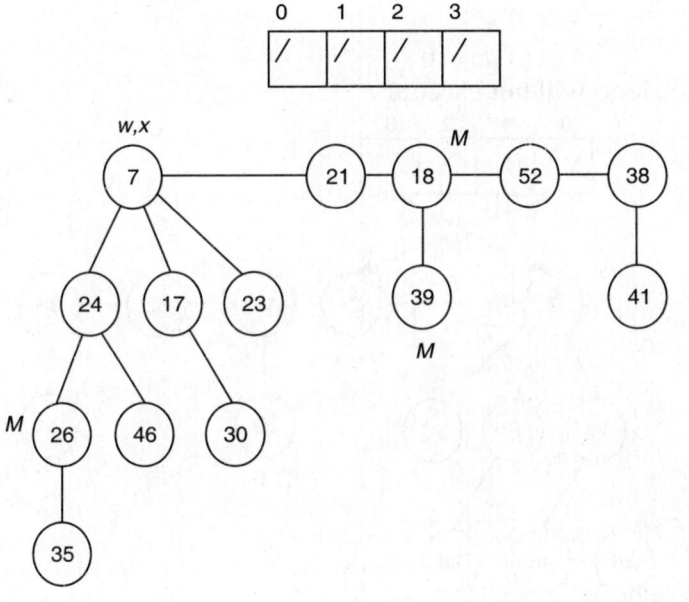

$$d = 3$$

Now while loop terminates

Following figure is obtained after the execution of Line 14

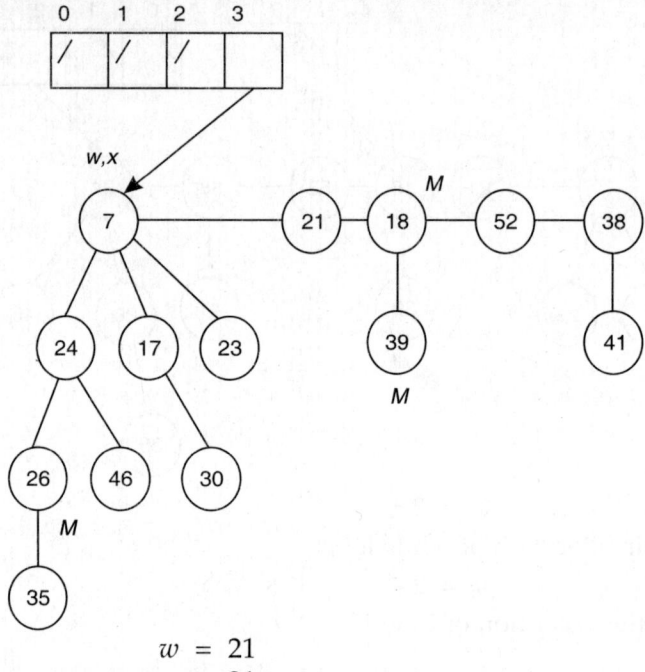

$$w = 21$$
$$x = 21$$
$$d = 0$$

while loop will not execute

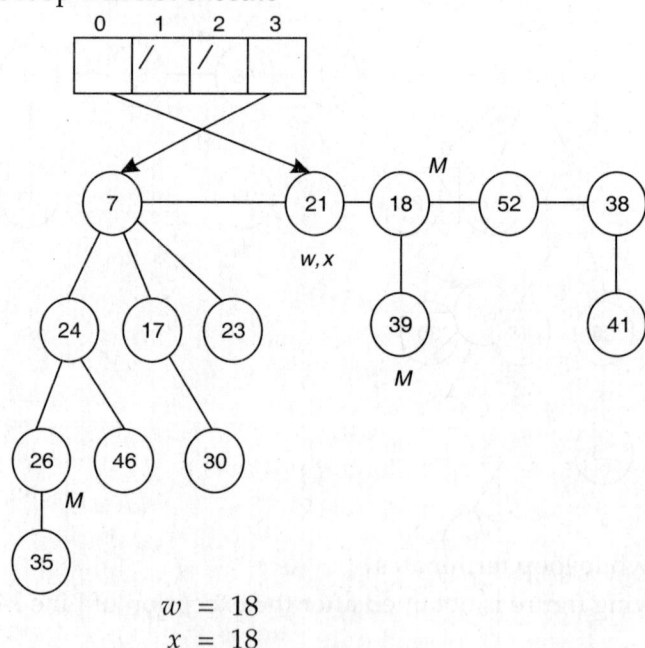

$$w = 18$$
$$x = 18$$
$$d = 1$$

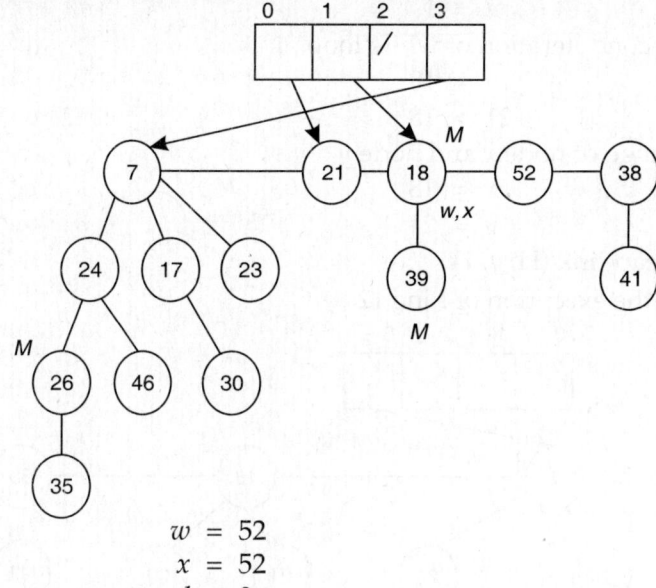

$$w = 52$$
$$x = 52$$
$$d = 0$$

while loop executes.

$$y = 21$$
$$52 > 21$$

exchange of node $x$ and node $y$

So,

$$x = 21$$
$$y = 52$$

Fib-heap-link (H, $y$, $x$)

After the execution of Line 12

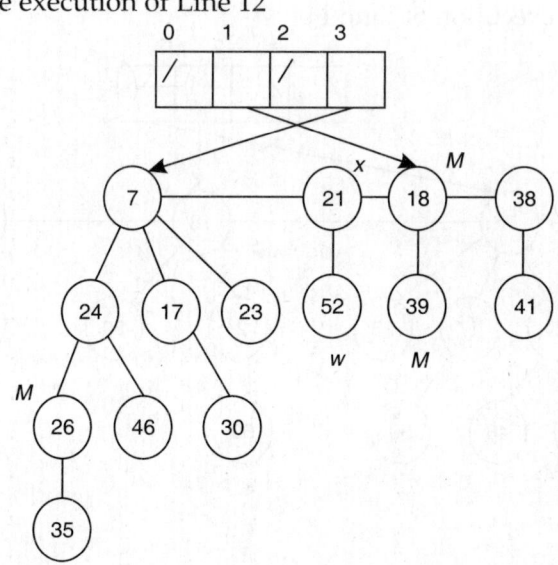

$$d = 1$$

For second iteration of while loop

$$y = 18$$
$$21 > 18$$

Exchange of node $x$ and node $y$

So,                      $x = 18$
$$y = 21$$

Fib-heap-link (H, $y$, $x$)

After the execution of Line 12

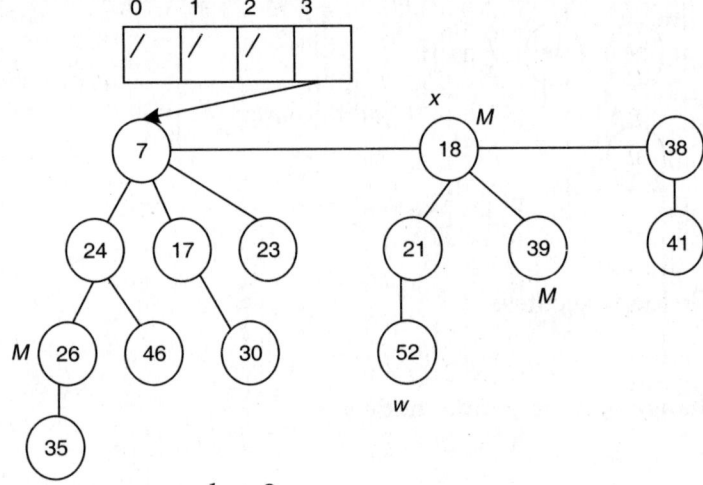

$$d = 2$$

while loop terminates

After the execution of Line 14

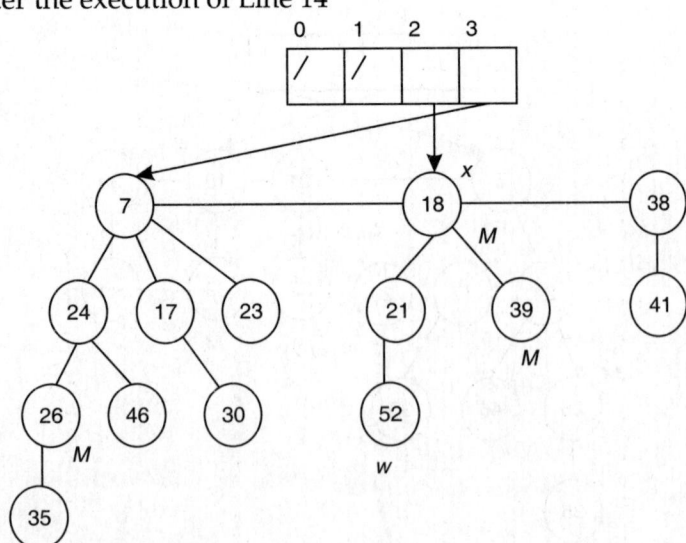

$$w = 38$$
$$x = 38$$
$$d = 1$$

while loop will not execute

Following figure is obtained after the execution of Line 14

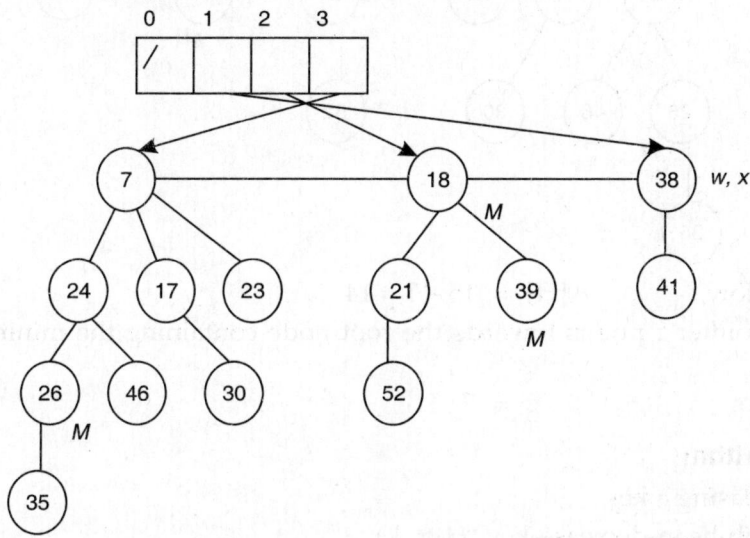

for loop of Line 4 terminates

          min [H] = nil, so root list becomes empty

for                i = 0

           A [0] = nil

for                i = 1

           A [1] ≠ nil

          min [H] = = nil

So, Root list of H contains 38

Now,       min [H] = 38

 for              i = 2

           A [2] ≠ nil

So, Root list of H contains 38, 18

            18 < 38

Now        min [H] = 18

for                i = 3

           A [3] ≠ nil

Root list of H contains 38, 18, 7

             7 < 18

Now        min [H] = 7

The for loop of Line 16 terminates

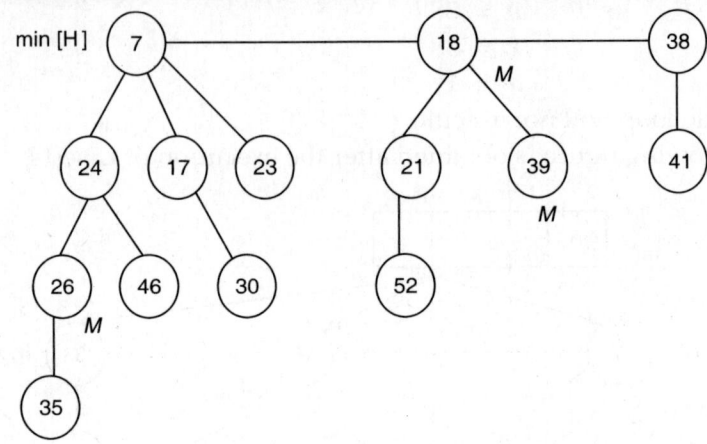

Now, $\qquad n\,[H]\;=\;15-1=14$

Pointer $z$ points towards the root node containing the minimum key

So, $\qquad\qquad z\;=\;7$

## Algorithm

Decreasing a key

    Fib-heap-decrease-key (H, x, k)

| | |
|---|---|
| Line 1 | if $\;k > key\,[x]$ |
| Line 2 | error "new key is greater than current key" |
| Line 3 | $key\,[x] \leftarrow k$ |
| Line 4 | $y \leftarrow p\,[x]$ |
| Line 5 | if $\;y \neq nil$ and $key\,[x] < key\,[y]$ |
| Line 6 | Cut (H, x, y) |
| Line 7 | Cascading-cut (H, y) |
| Line 8 | if $\;key\,[x] < key\,[min\,[H]]$ |
| Line 9 | $min\,[H] \leftarrow x$ |

## Explanation

- This procedure aims to assign a new key '$k$' to node $x$.
- Line 1 checks the *if* condition.
- Line 2 indicates that an error will be generated if the above *if* condition is true. The *if* condition is true when the new key value is greater than the current key value of $x$. In case of an error the algorithm will not proceed further. If the condition is false then the execution of Line 3 takes place.
- Line 3 indicates a new key $k$ is assigned to node $x$.

- Line 4 indicates that $y$ is the parent of $x$.
- Line 5 checks the *if* condition.
  This condition is true when both the cases are true.
  Case i.        $y \neq$ nil, i.e. $x$ is not the root node.
  Case ii.       key of $x$ is less than the key of its parent.
  When the *if* condition is true then the execution of Line 6 and Line 7 takes place.
- Line 6 calls Cut (H, $x$, $y$) procedure.
- Line 7 calls the procedure Cascading-cut (H, $y$).
- Line 8 checks the *if* condition. This condition is true when key [$x$] is greater than key [min [H]].
- Line 9 indicates that pointer min [H] is made to point to node $x$.

## Algorithm

Cut (H, x, y)

Line 1     remove $x$ from the child list of $y$, decrementing degree [y]

Line 2     add $x$ to the root list of H

Line 3     p [x] $\leftarrow$ nil

Line 4     mark [x] $\leftarrow$ false i.e. making the node x unmarked.

## Algorithm

Cascading-cut (H, $y$)

Line 1        $z \leftarrow p$ [y]

Line 2        *if*  $z \neq$ nil

Line 3            if · mark [y] = = false

Line 4                mark [y] $\leftarrow$ true

Line 5        *else*  Cut (H, $y$, $z$)

Line 6            Cascading-cut (H, $z$)

## Explanation

- Line 1 indicates that $z$ is made to point to the parent of node $y$.
- Line 2 checks the *if* condition. This condition is true when node y is not a root. If $y$ is a root then $z$ = = nil and the procedure will not proceed further.
- Line 3 checks the *if* condition. This condition is true when node $y$ is unmarked. If this condition fails, then the execution of Line 5 takes place.
- Line 4 marks node $y$.
- Line 5 calls the Cut (H, $y$, $z$) procedure.
- Line 6 calls the Cascading-cut (H, $z$).

## Analysis

- The Fib-heap-decrease-key procedure takes $O(1)$ time plus the time to perform Cascading cuts.
- Suppose that a given invocation of Fib-heap-decrease-key results in '$c$' calls of Cascading-cut (the call made from Line 7 of Fib-heap-decrease-key followed by $c - 1$ recursive calls of Cascading-cut). Each call of Cascading- cut takes $O(1)$ time exclusive of recursive calls. Therefore, the actual cost of Fib-heap-decrease-key, including all recursive calls is $O(c)$.
- The potential before calling the procedure Fib-heap-decrease-key is $t(H) + 2m(H)$.

  The call to Cut in Line 6 of Fib-heap-decrease-key creates a new tree rooted at node $x$ and clears $x$'s mark bit (which may have already been False). Each call of Cascading-cut, except for the last one, cuts a marked node and clears the mark bit.

  Afterward, the Fibonacci heap contains $t(H) + c$ trees (the original $t(H)$ trees, $c - 1$ trees produced by Cascading cuts, and the tree rooted at $x$) and at most $m(H) - c + 2$ marked nodes ($c - 1$ were unmarked by Cascading cuts and the last call of Cascading-cut may have marked a node).

  Thus, the change in potential is at most

  $$((t(H) + c) + 2(m(H) - c + 2)) - (t(H) + 2m(H)) = 4 - c$$

  Therefore, the amortized cost of Fib-heap-decrease-key is at most $O(c) + 4 - c = O(1)$ because we can scale up the units of potential to dominate the constant hidden in $O(c)$.

## Example

Apply Fib-heap-decrease-key $(H, x, k)$ operation on the following figure. Consider the node with key 46 and decrease its key to 15.

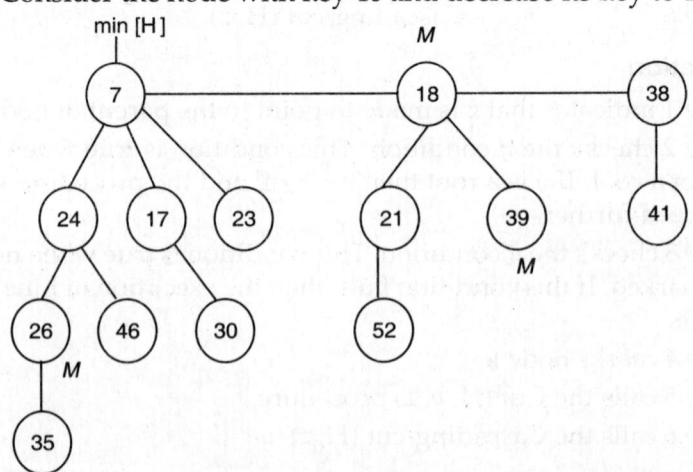

## Solution

x is the node with key value 46, i.e. key [x] = 46

$$15 \not> 46$$

Now          key [x] = 15

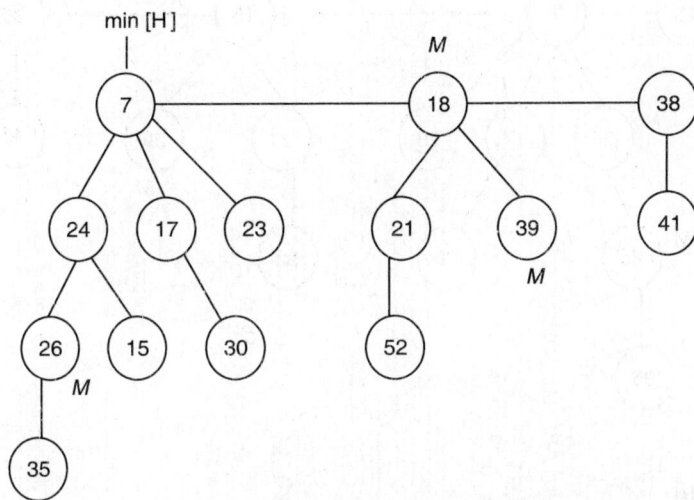

$y = 24$, i.e. $y$ is the node with key value 24.

$y \neq$ nil and $15 < 24$

Call procedure Cut (H, x, y)

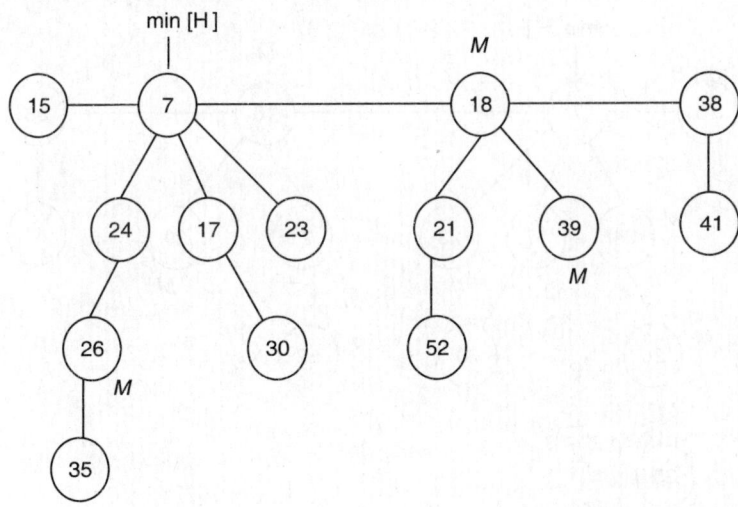

$$d [y] = 1$$
$$p [x] = \text{nil}$$

Call procedure Cascading-cut (H, y)

$z = 7$, i.e. $z$ is pointing towards the node with key 7

$z \neq nil$

Here mark $[y] = = $ false

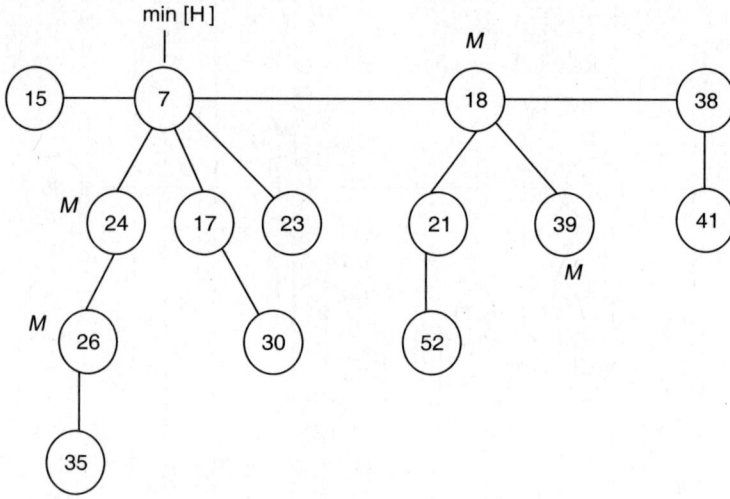

$$15 \nless 7$$

### Example

Apply Fib-heap-decrease-key (H, *x*, *k*) operation on the following figure.

Consider the node with key 35 and decrease its key to 5.

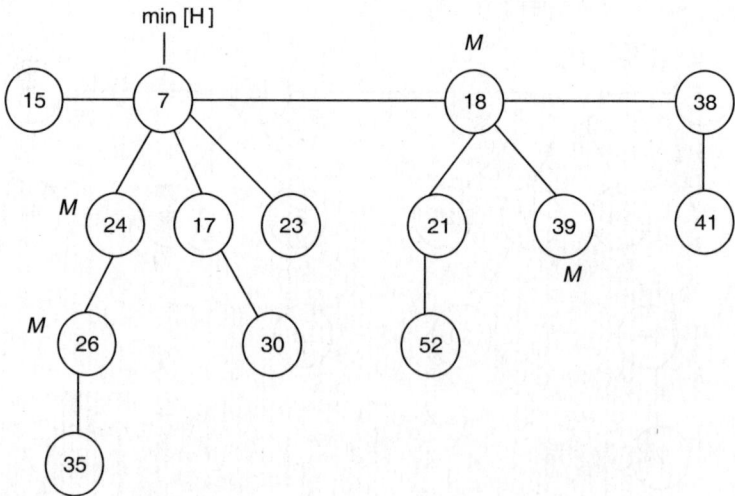

### Solution

*x* is the node with key value 35

$$k = 5$$

$$5 \ngtr 35$$

Now,        key $[x]$ = 5

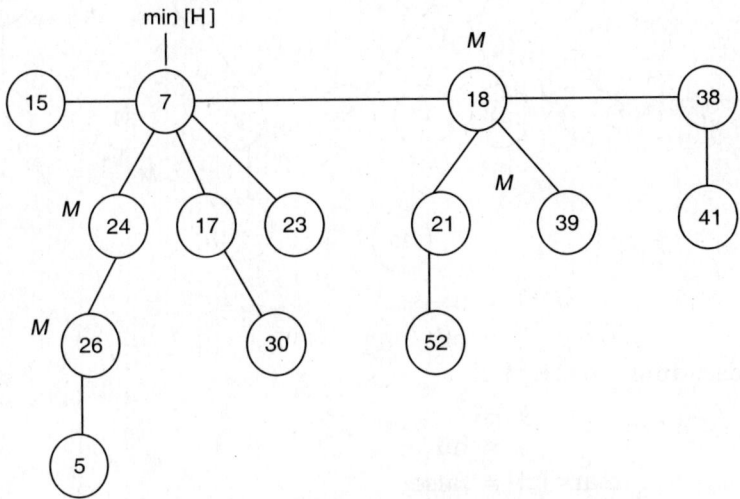

$y$ = 26, i.e. $y$ is the node with key 26.

$y \neq$ nil and 5 < 26

Call Cut (H, $x$, $y$) procedure

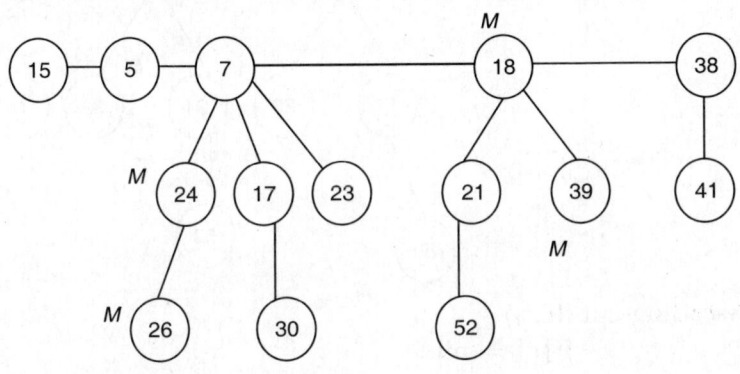

$d\,[y]$ = 0

Cascading-cut (H, $y$)

$z$ = 24

$z \neq$ nil

mark $[y] \neq$ false

Cut (H, $y$, $z$)

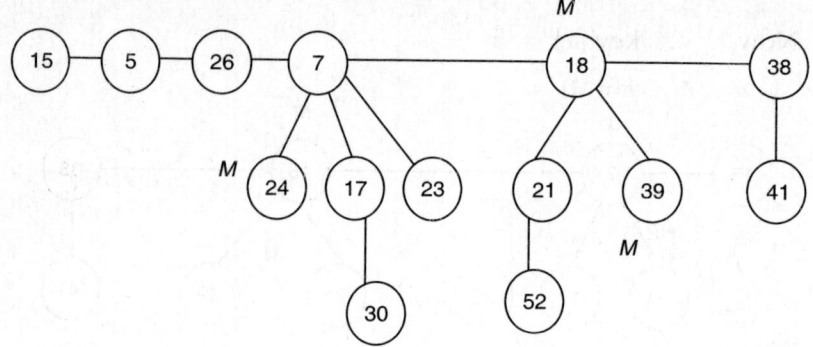

$$d\,[z] = 0$$
$$p\,[y] = \text{nil}$$
Cascading-cut (H, z)
$$y = 7$$
$$y \neq \text{nil}$$
$$\text{mark}\,[z] \neq \text{false}$$
Cut (H, z, y)

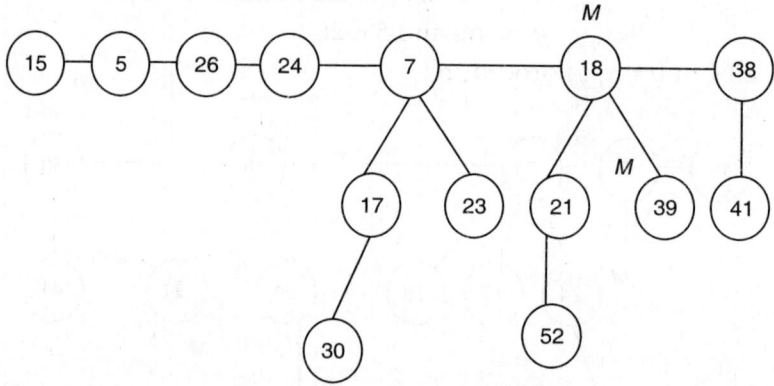

Cascading-cut (H, y)
$$p\,[y] = \text{nil}$$
$$z == \text{nil}$$
So, the procedure terminates
$$5 < 7$$
So,          min [H] = 5

## Algorithm

Deleting a node

   Fib-heap-delete (H, x)

Line 1          Fib-heap-decrease-key (H, x, − ∞)

Line 2          Fib-heap-extract-min (H)

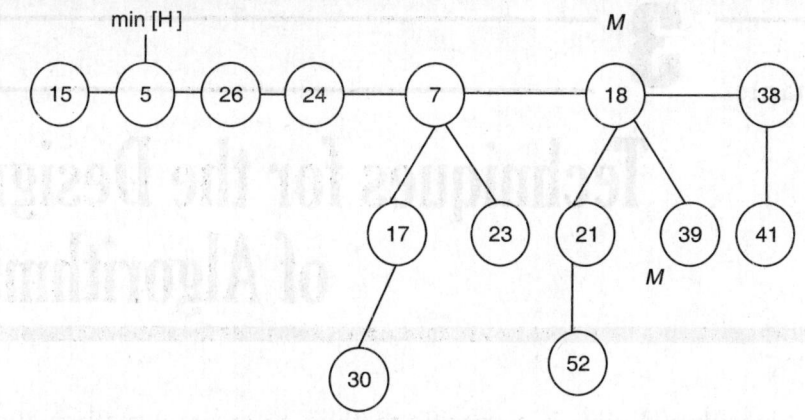

## Explanation

Fib-heap-delete procedure makes node $x$ to become the minimum node in the Fibonacci heap by giving it a uniquely small key of $-\infty$. The Fib-heap-extract-min procedure then removes node $x$ from the Fibonacci heap.

## Analysis

- The above procedure takes $O(D(n))$ amortized time.

# 2.8 DATA STRUCTURES FOR DISJOINT SETS

- A disjoint set data structure maintains a collection
  $$S = \{S_1, S_2, ..., S_k\} \text{ of disjoint dynamic sets.}$$
- Each set is identified by a 'representative' which is some member of the set.

## Disjoint-set Forests

- In a disjoint-set forest, each member points only to its parent. The root of each tree contains the representative and is its own parent.

## Heuristics to improve the running time

(i) *Union by rank*: In this heuristic, for each node we maintain a rank which is an upper bound on the height of the node. The root with smaller rank is made to point to the root with larger rank during a 'union' operation.

(ii) *Path compression*: This heuristic is used during 'Find-set' operations to make each node on the find path directly to the root. Path compression does not change any ranks.

# Chapter 3

# Techniques for the Design of Algorithms

- Algorithm design is a specific method to create a mathematical process in solving problems.
- The important aspect of algorithm design is to create an algorithm that has an efficient run time.

Some of the algorithm design approaches are:

1. Dynamic programming approach
2. Greedy approach
3. Divide and conquer approach
4. Decrease and conquer approach (incremental approach)
5. Data structure based approach
6. Backtracking approach

- Although more than one technique may be applicable to a specific problem, it is often the case that an algorithm constructed by one approach is clearly superior to equivalent solutions built using alternate techniques.

## 3.1 DYNAMIC PROGRAMMING APPROACH

- Dynamic programming approach is applied when the subproblems overlap.
- A dynamic programming algorithm solves each subproblems just once and then saves its answer in a table, hence avoiding the work of recomputing the answer every time it solves each subproblem.
- While developing a dynamic programming algorithm, following four steps have to be followed:
    (i) characterize the structure of the optimal solution
    (ii) recursively define the value of the optimal solution
    (iii) compute the value of the optimal solution, typically in a bottom up fashion.

(iv)  construct an optimal solution from computed information

## 3.1.1 Matrix Chain Multiplication

The matrix chain multiplication problem is stated as follows:

Given a chain $(A_1, A_2, A_3,...,A_n)$ of $n$ matrices, where $i = 1, 2, 3,...,n$, matrix $A_i$ has dimension $P_{i-1} \times P_i$, fully parenthesize the product $A_1 A_2...A_n$ in a way that minimizes the number of scalar multiplications.

In matrix chain multiplication problems, we are not actually multiplying the matrices. Here our aim is to only determine the order for multiplying matrices that has the lowest cost.

*Step 1: Structure of optimal parenthesization*

Consider the notation $A_{i...j}$, where $i \leq j$, for the matrix that results from evaluating the product $A_i A_{i+1}...A_j$.

If the problem is non-trivial, i.e. $i < j$, then to parenthesize the product $A_i A_{i+1}...A_j$, we must split the product between $A_k$ and $A_{k+1}$ for some integer $k$ in the range $i \leq k < j$. That is, for some value of $k$, we first compute the matrices $A_{i...k}$ and $A_{k+1...j}$ and then multiply them together to produce the final product $A_{i...j}$.

The cost of parenthesizing this way is the cost of computing the matrix $A_{i...k}$ plus the cost of computing $A_{k+1...j}$, plus the cost of multiplying them together.

*Step 2: A recursive solution*

Let $m[i, j]$ be the minimum number of scalar multiplications needed to compute the matrix $A_{i...j}$.

If the problem is trivial, i.e. $i = j$, the chain consists of just one matrix $A_{i...i} = A_i$, so that no scalar multiplications are necessary to compute the product. Hence $m[i, i] = 0$, where $i = 1, 2,...,n$.

If the problem is non-trivial, i.e. $i < j$, we split the product $A_i A_{i+1}...A_j$ between $A_k$ and $A_{k+1}$, where $i \leq k < j$. Then, $m[i, j]$ equals the minimum cost for computing the subproducts $A_{i...k}$ and $A_{k+1...j}$, plus the cost of multiplying these two matrices together. Thus, the recursive definition for the minimum cost of parenthesizing the product $A_i A_{i+1}...A_j$ is

$$m[i, j] = \begin{cases} 0 & \text{if } i = j \\ \min_{i \leq k < j} \left\{ m[i,k] + m[k+1,j] + P_{i-1} P_k P_j \right\} & \text{if } i < j \end{cases}$$

*Step 3: Computing the optimal costs*

The optimal cost is computed by using a tabular, bottom up approach.

Matrix-chain-order procedure is used to implement the tabular, bottom-up method.

## Algorithm

Matrix-chain-order (P)

Line 1        $n \leftarrow$ length [P] – 1
Line 2        let m [1...n, 1...n] and s [1...n – 1, 2...n] be new tables
Line 3        for i $\leftarrow$ 1 to n
Line 4        m [i, i] $\leftarrow$ 0
Line 5        for $\ell \leftarrow$ 2 to n
Line 6        for i $\leftarrow$ 1 to n – $\ell$ + 1
Line 7        j $\leftarrow$ i + $\ell$ – 1
Line 8        m [i, j] $\leftarrow \infty$
Line 9        for k $\leftarrow$ i to j – 1
Line 10       q $\leftarrow$ m [i, k] + m [k +1, j] + $P_{i-1} P_k P_j$
Line 11       if q < m [i, j]
Line 12       m [i, j] $\leftarrow$ q
Line 13       s [i, j] $\leftarrow$ k
Line 14       return m and s

## Explanation

This procedure assumes that matrix $A_i$ has dimensions $P_{i-1} \times P_i$ for $i = 1, 2, ..., n$. Its input is a sequence $P = \{P_0, P_1, ..., P_n\}$. The length of this sequence is stored in the attribute length [P].

- Line 1 indicates that the variable $n$ has the value one less than the length of sequence.
- Line 2 indicates that this procedure uses an auxiliary table m [1...n, 1...n] for storing the m [i, j] costs and another auxiliary table s [1...n –1, 2...n] that records which index of k achieved the optimal cost in computing m [i, j].
- Line 3 indicates the beginning of for loop that ends with Line 4. This for loop is applicable for i equals to one to n. The value of i is incremented by one after every iteration.
- Line 4 indicates that m [i, i] has been assigned a value zero for $i = 1, 2, ..., n$. This is the minimum cost for chains of length 1.
- Line 5 indicates the beginning of for loop that ends with Line 13. This loop runs for $\ell$ equals to 2 to n, where $\ell$ denotes the chain length. The value of $\ell$ is incremented by 1 after every iteration. Termination of this loop causes the control to go to Line 14.

- Line 6 indicates the beginning of for loop that ends with Line 13. It runs for $i$ equals to 1 to $n - \ell + 1$. The value of $i$ is incremented by 1 after every iteration. Termination of this loop causes the control to go back to Line 5.
- Line 7 indicates that the value of $j$ equals to $i + \ell - 1$.
- Line 8 indicates that the location $m[i, j]$ is set to $\infty$.
- Line 9 indicates the beginning of for loop that ends with Line 13. The variable $k$ takes values from $i$ to $j - 1$. The value of $k$ is incremented by 1 after every iteration. Termination of this loop causes the control to go back to Line 6.
- Line 10 indicates that $q$ has been assigned a value computed by adding $m[i, k]$, $m[k + 1, j]$ and the product of $P_{i-1}$, $P_k$ and $P_j$.
- Line 11 checks the *if* condition. This condition is true when the value of $q$ is less than the value at location $m[i, j]$ in the table. If this condition is true then the execution of Line 12 and Line 13 takes place else the control goes back to Line 9.
- Line 12 indicates that the value of $q$ is assigned to location $m[i, j]$.
- Line 13 indicates that the value of $k$ is assigned to location $s[i, j]$. Basically $s[i, j]$ is that value of $k$ at which we split the product $A_i A_{i+1}...A_j$ in an optimal parenthesization.
- Line 14 returns the $m$ and $s$ tables.

**Analysis**

- The algorithm Matrix-chain-order has a running time of $O(n^3)$.
- The loops are nested three deep and each loop index ($\ell$, $i$ and $k$) takes on at most $n - 1$ values.

  The algorithm requires $\Theta(n^2)$ space to store $m$ and $s$ tables.

*Step 4: Constructing an optimal solution.*

An optimal parenthesization of $(A_i A_{i+1}...A_j)$ is obtained through Print-optimal-parens $(s, i, j)$ procedure, given the $s$ table computed by Matrix-chain-order $(P)$ procedure and the indices $i$ and $j$.

**Algorithm**

```
Print-optimal-parens (s, i, j)
Line 1      if i = = j
Line 2          print "Aᵢ"
Line 3      else print "("
Line 4          Print-optimal-parens (s, i, s [i, j])
Line 5          Print-optimal-parens (s, s [i, j] + 1, j)
Line 6          print ")"
```

## Explanation

- Line 1 checks the *if* condition. This condition is true if $i$ equals to $j$. If the condition is true then the execution of Line 2 takes place. If the condition is false then the execution from Line 3 to Line 6 takes place.
- Line 2 prints $A_i$
- Line 3 prints (
- Line 4 calls the procedure Print-optimal-parens $(s, i, s\,[i, j])$.
- Line 5 calls the procedure Print-optimal-parens $(s, s\,[i, j] + 1, j)$.
- Line 6 prints )

## Example

Find an optimal parenthesization of a matrix chain product of the sequence [5, 4, 2, 6, 7].

## Solution

$$P = <5, 4, 2, 6, 7>$$
$$P_0 = 5$$
$$P_1 = 4$$
$$P_2 = 2$$
$$P_3 = 6$$
$$P_4 = 7$$

$A_1 = 5 \times 4,\ A_2 = 4 \times 2,\ A_3 = 2 \times 6,\ A_4 = 6 \times 7$

length $[P] = 5$

$n = 5 - 1 = 4$

After the execution of for loop of Line 3

| 1 | 2 | 3 | 4 | j/i |
|---|---|---|---|---|
| 0 |  |  |  | 1 |
|  | 0 |  |  | 2 |
|  |  | 0 |  | 3 |
| $m$ |  |  | 0 | 4 |

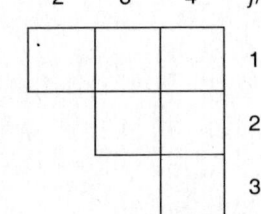

| 2 | 3 | 4 | j/i |
|---|---|---|---|
| . |  |  | 1 |
|  |  |  | 2 |
| $s$ |  |  | 3 |

$$\ell = 2$$

$i$ takes values from 1 to $4 - 2 + 1$

$$i = 1$$
$$j = 1 + 2 - 1$$
$$\therefore \qquad j = 2$$

| 1 | 2 | 3 | 4 | j/i |
|---|---|---|---|---|
| 0 | $\infty$ |  |  | 1 |
|  | 0 |  |  | 2 |
|  |  | 0 |  | 3 |
|  | $m$ |  | 0 | 4 |

$k$ can take values from 1 to $2 - 1$

$$k = 1$$
$$q = m\,[1, 1] + m\,[2, 2] + P_0\,P_1\,P_2$$
$$q = 0 + 0 + 5 \times 4 \times 2$$
$$q = 40$$
$$40 < \infty$$

| 1 | 2 | 3 | 4 | j/i |
|---|---|---|---|---|
| 0 | 40 |  |  | 1 |
|  | 0 |  |  | 2 |
|  | m | 0 |  | 3 |
|  |  |  | 0 | 4 |

| 2 | 3 | 4 | j/i |
|---|---|---|---|
| 1 |  |  | 1 |
|  |  |  | 2 |
| s |  |  | 3 |

$$i = 2$$
$$j = 2 + 2 - 1$$
$$\therefore \quad j = 3$$
$$m\,[2, 3] = \infty$$

| 1 | 2 | 3 | 4 | j/i |
|---|---|---|---|---|
| 0 | 40 |  |  | 1 |
|  | 0 | ∞ |  | 2 |
|  |  | 0 |  | 3 |
|  | m |  | 0 | 4 |

$$k = 2 \text{ to } 3 - 1$$
$$q = m\,[2, 2] + m\,[3, 3] + P_1 P_2 P_3$$
$$q = 0 + 0 + 48$$
$$q = 48$$
$$48 < \infty$$

| 1 | 2 | 3 | 4 | j/i |
|---|---|---|---|-----|
| 0 | 40 |  |  | 1 |
|  | 0 | 48 |  | 2 |
|  |  | 0 |  | 3 |
| m |  |  | 0 | 4 |

| 2 | 3 | 4 | j/i |
|---|---|---|-----|
| 1 |  |  | 1 |
|  | 2 |  | 2 |
| s |  |  | 3 |

$$i = 3$$
$$j = 3 + 2 - 1$$
$$j = 4$$

| 1 | 2 | 3 | 4 | j/i |
|---|---|---|---|-----|
| 0 | 40 |  |  | 1 |
|  | 0 | 48 |  | 2 |
|  |  | 0 | $\infty$ | 3 |
| m |  |  | 0 | 4 |

$$k = 3 \text{ to } 4 - 1$$
$$q = m[3, 3] + m[4, 4] + P_2 P_3 P_4$$
$$q = 84$$
$$84 < \infty$$

| 1 | 2 | 3 | 4 | j/i |
|---|---|---|---|---|
| 0 | 40 |   |   | 1 |
|   | 0 | 48 |   | 2 |
|   |   | 0 | 84 | 3 |
|   | m |   | 0 | 4 |

| 2 | 3 | 4 | j/i |
|---|---|---|---|
| 1 |   |   | 1 |
|   | 2 |   | 2 |
| s |   | 3 | 3 |

$$\ell = 3$$

$i$ takes values from 1 to $4 - 3 + 1$

$$i = 1$$
$$j = 1 + 3 - 1$$
$$\therefore \quad j = 3$$

| 1 | 2 | 3 | 4 | j/i |
|---|---|---|---|---|
| 0 | 40 | ∞ |   | 1 |
|   | 0 | 48 |   | 2 |
|   |   | 0 | 84 | 3 |
|   | m |   | 0 | 4 |

$k$ takes values from 1 to $3 - 1$

$$k = 1$$
$$q = m[1, 1] + m[2, 3] + P_0 P_1 P_3$$
$$q = 0 + 48 + 120$$
$$q = 168$$

| 1 | 2 | 3 | 4 | j/i |
|---|----|-----|----|---|
| 0 | 40 | 168 |    | 1 |
|   | 0  | 48  |    | 2 |
|   |    | 0   | 84 | 3 |
|   | m  |     | 0  | 4 |

| 2 | 3 | 4 | j/i |
|---|---|---|---|
| 1 | 1 |   | 1 |
|   | 2 |   | 2 |
| s |   | 3 | 3 |

$$k = 2$$
$$q = m[1, 2] + m[3, 3] + P_0 P_2 P_3$$
$$q = 40 + 0 + 60$$
$$q = 100$$
$$100 < 168$$

| 1 | 2 | 3 | 4 | j/i |
|---|----|-----|----|---|
| 0 | 40 | 100 |    | 1 |
|   | 0  | 48  |    | 2 |
|   |    | 0   | 84 | 3 |
|   | m  |     | 0  | 4 |

| 2 | 3 | 4 | j/i |
|---|---|---|---|
| 1 | 2 |   | 1 |
|   | 2 |   | 2 |
| s |   | 3 | 3 |

$$i = 2$$
$$j = 2 + 3 - 1$$
$$\therefore \qquad j = 4$$
$$m\,[2, 4] = \infty$$

| 1 | 2 | 3 | 4 | j/i |
|---|---|---|---|---|
| 0 | 40 | 100 | | 1 |
| | 0 | 48 | ∞ | 2 |
| | | 0 | 84 | 3 |
| | m | | 0 | 4 |

$$k = 2 \text{ to } 4 - 1$$
$$k = 2$$
$$q = m\,[2, 2] + m\,[3, 4] + P_1\,P_2\,P_4$$
$$q = 0 + 84 + (4 \times 2 \times 7)$$
$$q = 84 + 56$$
$$q = 140$$
$$140 < \infty$$

| 1 | 2 | 3 | 4 | j/i |
|---|---|---|---|---|
| 0 | 40 | 100 | | 1 |
| | 0 | 48 | 140 | 2 |
| | | 0 | 84 | 3 |
| | m | | 0 | 4 |

| 2 | 3 | 4 | j/i |
|---|---|---|---|
| 1 | 2 | | 1 |
| | 2 | 2 | 2 |
| s | | 3 | 3 |

$$k = 3$$
$$q = m[2, 3] + m[4, 4] + P_1 P_3 P_4$$
$$q = 48 + 0 + (4 \times 6 \times 7)$$
$$q = 216$$
$$216 \not< 140$$
$$\ell = 4$$

$i$ takes values from 1 to $4 - 4 + 1$

$$i = 1$$
$$j = 1 + 4 - 1$$
$$\therefore \qquad j = 4$$
$$m[1, 4] = \infty$$

| 1 | 2 | 3 | 4 | j/i |
|---|---|---|---|-----|
| 0 | 40 | 100 | ∞ | 1 |
|  | 0 | 48 | 140 | 2 |
|  |  | 0 | 84 | 3 |
|  | m |  | 0 | 4 |

$$k = 1 \text{ to } 4 - 1$$
$$k = 1$$
$$q = m[1, 1] + m[2, 4] + P_0 P_1 P_4$$
$$q = 0 + 140 + (5 \times 4 \times 7)$$
$$q = 280$$
$$280 < \infty$$

| 1 | 2 | 3 | 4 | j/i |
|---|---|---|---|-----|
| 0 | 40 | 100 | 280 | 1 |
|  | 0 | 48 | 140 | 2 |
|  |  | 0 | 84 | 3 |
|  | m |  | 0 | 4 |

| 2 | 3 | 4 | j/i |
|---|---|---|-----|
| 1 | 2 | 1 | 1 |
|   | 2 | 2 | 2 |
| s |   | 3 | 3 |

$$k = 2$$
$$q = m[1, 2] + m[3, 4] + P_0 P_2 P_4$$
$$= 40 + 84 + (5 \times 2 \times 7)$$
$$= 194$$
$$194 < 280$$

| 1 | 2 | 3 | 4 | j/i |
|---|---|---|---|-----|
| 0 | 40 | 100 | 194 | 1 |
|   | 0 | 48 | 140 | 2 |
|   |   | 0 | 84 | 3 |
|   | m |   | 0 | 4 |

| 2 | 3 | 4 | j/i |
|---|---|---|-----|
| 1 | 2 | 2 | 1 |
|   | 2 | 2 | 2 |
| s |   | 3 | 3 |

$$k = 3$$
$$q = m[1, 3] + m[4, 4] + P_0 P_3 P_4$$
$$q = 100 + 0 + (5 \times 6 \times 7)$$
$$q = 100 + 210$$
$$q = 310$$
$$194 \not< 310$$

[After the execution of Line 14 of procedure Matrix-chain-order $(P)$]

| 1 | 2 | 3 | 4 | j/i |
|---|---|---|---|---|
| 0 | 40 | 100 | 194 | 1 |
|   | 0 | 48 | 140 | 2 |
|   |   | 0 | 84 | 3 |
|   |   |   | 0 | 4 |

m

| 2 | 3 | 4 | j/i |
|---|---|---|---|
| 1 | 2 | 2 | 1 |
|   | 2 | 2 | 2 |
|   |   | 3 | 3 |

s

Print-optimal-parens $(s, 1, 4)$

$1 \neq 4$

(

Print-optimal-parens $(s, 1, s\,[1, 4])$ = Print-optimal-parens $(s, 1, 2)$

$1 \neq 2$

(

Print-optimal-parens $(s, 1, s\,[1, 2])$ = Print-optimal-parens $(s, 1, 1)$

$1 == 1$

$A_1$

Print-optimal-parens $(s, s\,[1, 2] + 1, j)$ = Print-optimal-parens $(s, 2, 2)$

$2 == 2$

$A_2$

)

Print-optimal-parens $(s, s\,[1, 4] + 1, j)$ = Print-optimal-parens $(s, 3, 4)$

$3 \neq 4$

(

Print-optimal-parens $(s, 3, s\,[3, 4])$ = Print-optimal-parens $(s, 3, 3)$

$3 == 3$

$A_3$

Print-optimal-parens $(s, s\,[3, 4] + 1, 4)$ = Print-optimal-parens $(s, 4, 4)$

$4 = = 4$

$A_4$

)

)

Thus, the Print-optimal-parens $(s, 1, 4)$ prints the parenthesization $((A_1\,A_2)\,(A_3\,A_4))$

## Example

Find an optimal parenthesization of a matrix chain product of the sequence $[5, 6, 8, 4, 3, 2, 4, 7]$

## Solution

$$P = [5, 6, 8, 4, 3, 2, 4, 7]$$
$$P_0 = 5$$
$$P_1 = 6$$
$$P_2 = 8$$
$$P_3 = 4$$
$$P_4 = 3$$
$$P_5 = 2$$
$$P_6 = 4$$
$$P_7 = 7$$
$$A_1 = 5 \times 6,\ A_2 = 6 \times 8,\ A_3 = 8 \times 4,\ A_4 = 4 \times 3,$$
$$A_5 = 3 \times 2,\ A_6 = 2 \times 4,\ A_7 = 4 \times 7$$
$$\text{length}\,[P] = 8$$
$$n = 8 - 1 = 7$$
$$m\,[1, 2] = m\,[1,1] + m\,[2, 2] + P_0\,P_1 P_2 = 240$$
$$m\,[2, 3] = m\,[2,2] + m\,[3, 3] + P_1\,P_2 P_3 = 192$$
$$m\,[3, 4] = m\,[3,3] + m\,[4, 4] + P_2\,P_3 P_4 = 96$$
$$m\,[4, 5] = m\,[4,4] + m\,[5, 5] + P_3\,P_4 P_5 = 24$$
$$m\,[5, 6] = m\,[5,5] + m\,[6, 6] + P_4\,P_5 P_6 = 24$$
$$m\,[6, 7] = m\,[6,6] + m\,[7, 7] + P_5\,P_6 P_7 = 56$$

$$m\,[1,3] = \min \begin{cases} m[1,1] + m[2,3] + P_0 P_1 P_3 \\ m[1,2] + m[3,3] + P_0 P_2 P_3 \end{cases}$$

$$= \min \begin{cases} 312 \\ 400 \end{cases}$$

$$= 312$$

$$m\,[2,4] \;=\; \min \begin{cases} m[2,2]+m[3,4]+P_1P_2P_4 \\ m[2,3]+m[4,4]+P_1P_3P_4 \end{cases}$$

$$=\; \min \begin{cases} 240 \\ 264 \end{cases}$$

$$=\; 240$$

$$m\,[3,5] \;=\; \min \begin{cases} m[3,3]+m[4,5]+P_2P_3P_5 \\ m[3,4]+m[5,5]+P_2P_4P_5 \end{cases}$$

$$=\; \min \begin{cases} 88 \\ 144 \end{cases}$$

$$=\; 88$$

$$m\,[4,6] \;=\; \min \begin{cases} m[4,4]+m[5,6]+P_3P_4P_6 \\ m[4,5]+m[6,6]+P_3P_5P_6 \end{cases}$$

$$=\; \min \begin{cases} 72 \\ 56 \end{cases}$$

$$=\; 56$$

$$m\,[5,7] \;=\; \min \begin{cases} m[5,5]+m[6,7]+P_4P_5P_7 \\ m[5,6]+m[7,7]+P_4P_6P_7 \end{cases}$$

$$=\; \min \begin{cases} 98 \\ 108 \end{cases}$$

$$=\; 98$$

$$m\,[1,4] \;=\; \min \begin{cases} m[1,1]+m[2,4]+P_0P_1P_4 \\ m[1,2]+m[3,4]+P_0P_2P_4 \\ m[1,3]+m[4,4]+P_0P_3P_4 \end{cases}$$

$$=\; \min \begin{cases} 330 \\ 456 \\ 372 \end{cases}$$

$$=\; 330$$

$$m\,[2,5] \;=\; \min \begin{cases} m[2,2]+m[3,5]+P_1P_2P_5 \\ m[2,3]+m[4,5]+P_1P_3P_5 \\ m[2,4]+m[5,5]+P_1P_4P_5 \end{cases}$$

$$= \min \begin{cases} 184 \\ 264 \\ 276 \end{cases}$$

$$= 184$$

$$m\,[3,6] = \min \begin{cases} m[3,3] + m[4,6] + P_2 P_3 P_6 \\ m[3,4] + m[5,6] + P_2 P_4 P_6 \\ m[3,5] + m[6,6] + P_2 P_5 P_6 \end{cases}$$

$$= \min \begin{cases} 184 \\ 216 \\ 152 \end{cases}$$

$$= 152$$

$$m\,[4,7] = \min \begin{cases} m[4,4] + m[5,7] + P_3 P_4 P_7 \\ m[4,5] + m[6,7] + P_3 P_5 P_7 \\ m[4,6] + m[7,7] + P_3 P_6 P_7 \end{cases}$$

$$= \min \begin{cases} 182 \\ 136 \\ 168 \end{cases}$$

$$= 136$$

$$m\,[1,5] = \min \begin{cases} m[1,1] + m[2,5] + P_0 P_1 P_5 \\ m[1,2] + m[3,5] + P_0 P_2 P_5 \\ m[1,3] + m[4,5] + P_0 P_3 P_5 \\ m[1,4] + m[5,5] + P_0 P_4 P_5 \end{cases}$$

$$= \min \begin{cases} 244 \\ 408 \\ 376 \\ 360 \end{cases}$$

$$= 244$$

$$m\,[2,6] = \min \begin{cases} m[2,2] + m[3,6] + P_1 P_2 P_6 \\ m[2,3] + m[4,6] + P_1 P_3 P_6 \\ m[2,4] + m[5,6] + P_1 P_4 P_6 \\ m[2,5] + m[6,6] + P_1 P_5 P_6 \end{cases}$$

$$= \min \begin{cases} 344 \\ 344 \\ 336 \\ 230 \end{cases}$$

$$= 230$$

$$m[3,7] = \min \begin{cases} m[3,3] + m[4,7] + P_2 P_3 P_7 \\ m[3,4] + m[5,7] + P_2 P_4 P_7 \\ m[3,5] + m[6,7] + P_2 P_5 P_7 \\ m[3,6] + m[7,7] + P_2 P_6 P_7 \end{cases}$$

$$= \min \begin{cases} 360 \\ 362 \\ 256 \\ 376 \end{cases}$$

$$= 256$$

$$m[1,6] = \min \begin{cases} m[1,1] + m[2,6] + P_0 P_1 P_6 \\ m[1,2] + m[3,6] + P_0 P_2 P_6 \\ m[1,3] + m[4,6] + P_0 P_3 P_6 \\ m[1,4] + m[5,6] + P_0 P_4 P_6 \\ m[1,5] + m[6,6] + P_0 P_5 P_6 \end{cases}$$

$$= \min \begin{cases} 350 \\ 552 \\ 448 \\ 414 \\ 284 \end{cases}$$

$$= 284$$

$$m[2,7] = \min \begin{cases} m[2,2] + m[3,7] + P_1 P_2 P_7 \\ m[2,3] + m[4,7] + P_1 P_3 P_7 \\ m[2,4] + m[5,7] + P_1 P_4 P_7 \\ m[2,5] + m[6,7] + P_1 P_5 P_7 \\ m[2,6] + m[7,7] + P_1 P_6 P_7 \end{cases}$$

$$= \min \begin{cases} 592 \\ 498 \\ 464 \\ 324 \\ 398 \end{cases}$$

$$= 324$$

$$m\,[1,7] \;=\; \min \begin{cases} m[1,1] + m[2,7] + P_0 P_1 P_7 \\ m[1,2] + m[3,7] + P_0 P_2 P_7 \\ m[1,3] + m[4,7] + P_0 P_3 P_7 \\ m[1,4] + m[5,7] + P_0 P_4 P_7 \\ m[1,5] + m[6,7] + P_0 P_5 P_7 \\ m[1,6] + m[7,7] + P_0 P_6 P_7 \end{cases}$$

$$= \min \begin{cases} 534 \\ 576 \\ 588 \\ 533 \\ 370 \\ 424 \end{cases}$$

$$= 370$$

| 1 | 2 | 3 | 4 | 5 | 6 | 7 | j/i |
|---|---|---|---|---|---|---|-----|
| 0 | 240 | 312 | 330 | 244 | 284 | 370 | 1 |
|   | 0 | 192 | 240 | 184 | 230 | 324 | 2 |
|   |   | 0 | 96 | 88 | 152 | 256 | 3 |
|   |   |   | 0 | 24 | 56 | 136 | 4 |
|   |   |   |   | 0 | 24 | 98 | 5 |
|   |   |   |   |   | 0 | 56 | 6 |
|   |   |   | m |   |   | 0 | 7 |

| 2 | 3 | 4 | 5 | 6 | 7 | j/i |
|---|---|---|---|---|---|---|
| 1 | 1 | 1 | 1 | 5 | 5 | 1 |
|   | 2 | 2 | 2 | 5 | 5 | 2 |
|   |   | 3 | 3 | 5 | 5 | 3 |
|   |   |   | 4 | 5 | 5 | 4 |
|   |   |   |   | 5 | 5 | 5 |
|   |   |   | s |   | 6 | 6 |

The Print-optimal-parens $(s, 1, 7)$ prints the prenthesization

$((A_1 (A_2 (A_3 (A_4 A_5)))) (A_6 A_7))$

## 3.1.2 Longest Common Subsequence (LCS)

Given two sequences $X$ and $Y$, we say that a sequence $Z$ of length $L$ is a longest common subsequence of $X$ and $Y$, if $Z$ is a subsequence of both X and Y and there exists no other common subsequence of length greater than $L$:

e.g.              $X = [A, B, C, B, D]$
                  $Y = [A, C, B, D, E]$
then,             $Z = [A, C, B, D]$
$Z$ is the LCS of $X$ and $Y$

### Problem

In the longest common subsequence problem, we are given two sequences

$X = [x_1, x_2,..., x_m]$ and $Y = [y_1, y_2,..., y_n]$ and wish to find a maximum length common subsequence of $X$ and $Y$.

**Step 1:** *Characterizing a longest common subsequence*

Given a sequence $X = [x_1, x_2, x_3, ..., x_m]$, we define the $i$th 'prefix' of X, where $i = 0, 1, 2, ..., m$, as $X_i = [x_1, x_2, ..., x_i]$

e.g.              if $X = [A, D, B, C, E]$
then              $X_4 = [A, D, B, C]$

Following theorem characterizes the longest common subsequences and states that an LCS of two sequences contains, within it, an LCS of prefixes of the two sequences. Thus, the LCS problem has an optimal substructure property.

*Theorem*: (optimal structure of an LCS)

Let $X = [x_1, x_2, ..., x_m]$ and $Y = [y_1, y_2, ..., y_n]$ be sequences, and let $Z = [z_1, z_2, ..., z_k]$ be any LCS of $x$ and $y$.

1. If $x_m = y_n$, then $z_k = x_m = y_n$ and $Z_{k-1}$ is an LCS of $X_{m-1}$ and $Y_{n-1}$
2. If $x_m \neq y_n$, then $z_k \neq x_m$ implies that $Z$ is an LCS of $X_{m-1}$ and $Y$
3. If $x_m \neq y_n$, then $z_k \neq y_n$ implies that $Z$ is an LCS of $X$ and $Y_{n-1}$

The above theorem implies that we should examine either one or two subproblems when finding an LCS of $X = [x_1, x_2, ..., x_m]$ and $Y = [y_1, y_2, ..., y_n]$.

If $x_m = y_n$, we must find an LCS of $X_{m-1}$ and $Y_{n-1}$. Appending $x_m = y_n$ to this LCS yields an LCS of $X$ and $Y$.

If $x_m \neq y_n$, then we must solve two subproblems

(i) Finding an LCS of $X_{m-1}$ and $Y$

(ii) Finding an LCS of $X$ and $Y_{n-1}$

**Step 2:** *A recursive solution*

Let $c[i, j]$ be the length of an LCS of the sequences $X_i$ and $Y_j$

The optimal substructure of the LCS problem gives the following recursive formula:

$$c[i, j] = \begin{cases} 0 & if\ i = 0 \quad or \quad j = 0 \\ c[i-1, j-1]+1 & if\ i, j > 0 \quad and \quad x_i = y_j \\ \max\begin{cases} c[i, j-1] \\ c[i-1, j] \end{cases} & if\ i, j > 0 \ and \quad x_i \neq y_j \end{cases}$$

If, either $i = 0$ or $j = 0$, one of the sequences has length 0, and so the LCS has length 0. When $x_i = y_j$, we consider the problem of finding an LCS of $X_{i-1}$ and $Y_{j-1}$.

Otherwise we consider the two subproblems of finding an LCS of $X_i$ and $Y_{j-1}$ and of $X_{i-1}$ and $Y_j$.

**Step 3:** *Computing the length of an LCS*

The following procedure computes the length of an LCS.

## Algorithm

LCS-Length (X, Y)

Line 1      m ← length [X]

Line 2      n ← length [Y]

Line 3      Let b [1 ... m, 1 ... n] and c [0...m, 0 ...n] be new tables

Line 4      for i ← 1 to m

Line 5          c [i, 0] ← 0
Line 6     for j ← 0 to n
Line 7          c [0, j] ← 0
Line 8     for i ← 1 to m
Line 9          for j ← 1 to n
Line 10             if $x_i$ = = $y_j$
Line 11                 c [i, j] ← c [i – 1, j – 1] + 1
Line 12                 b [i, j] ← " ↖ "
Line 13             elseif c [i – 1, j] ≥ c [i, j – 1]
Line 14                 c [i, j] ← c [i – 1, j]
Line 15                 b [i, j] ← "↑"
Line 16             else c [i, j] ← c [i, j – 1]
Line 17                 b [i, j] ← "←"
Line 18     return c and b

## Explanation

- The procedure LCS-Length takes two sequences $X = [x_1, x_2, ..., x_m]$ and $Y = [y_1, y_2, ..., y_n]$ as inputs.
- Line 1 indicates that the variable $m$ stores the length of sequence $X$.
- Line 2 indicates that the variable $n$ stores the length of sequence $Y$.
- Line 3 indicates that the procedure maintains the table $b$ [1...$m$, 1...$n$] which helps in constructing an optimal solution. The procedure stores the $c$ [$i, j$] values in a table $c$ [0, ..., $m$, 0, ..., $n$], and it fills in the first row of $c$ from left to right and then the second row and so on.
- Line 4 indicates the beginning of for loop that ends with Line 5. This loop is applicable for $i$ equals to 1 to $m$. Each time the value of $i$ is incremented by 1.
- Line 5 indicates that the location $c$ [$i, 0$] is set to zero.
- Line 6 indicates the beginning of for loop that ends with Line 7. This loop is applicable for $i$ equals to 0 to $n$. The value of $i$ is incremented by 1 after every iteration.
- Line 7 indicates that the location $c$ [$0, j$] is set to zero.
- Line 8 indicates the beginning of for loop that ends with Line 17. This loop is applicable for $i$ equals to 1 to $m$. The value of $i$ is incremented by 1 after every iteration.
- Line 9 indicates the beginning of for loop that ends with Line 17. This loop is applicable for $j$ equals to 1 to $n$. $j$ is incremented by 1 after every iteration. Termination of this for loop causes the control to go back to Line 8.

- Line 10 checks the *if* condition. This condition is true when $x_i$ equals to $y_j$. If this condition is true then the execution of Lines 11 and 12 takes place otherwise the control goes to Line 13.
- Line 11 indicates that the location $c[i, j]$ is set to a value which is one greater than the value at location $c[i-1, j-1]$.
- Line 12 indicates that ↖ sign is put at location $b[i, j]$.
- Line 13 checks the elseif condition. This condition is true if the value at location $c[i-1, j]$ is greater than or equal to the value at location $c[i, j-1]$. If this condition is true then the execution of Line 14 and Line 15 takes place else the execution of Line 16 and Line 17 takes place.
- Line 14 indicates that the location $c[i, j]$ is set to the value at location $c[i-1, j]$.
- Line 15 indicates that ↑ sign is put at location $b[i, j]$.
- Line 16 indicates that $c[i, j]$ is set to a value at location $c[i, j-1]$.
- Line 17 indicates that ← is placed at location $b[i, j]$.
- Line 18 returns tables $c$ and $b$.

### Analysis

- The running time of the procedure is $\Theta(m, n)$ because each table entry takes $\Theta(1)$ time to compute.

*Step* **4**: *Constructing an LCS*

The $b$ table returned by LCS-Length enables us to construct an LCS of

$$X = [x_1, x_2, ..., x_m] \text{ and } Y = [y_1, y_2, ..., y_n].$$

We begin at $b[m, n]$ and trace through the table by following the arrows.

Whenever, we encounter ↖ in entry $b[i, j]$, it implies that $x_i = y_j$ is an element of the LCS, that LCS-Length found. With this method, we encounter the elements of this LCS in reverse order.

### Example

Determine an LCS of the sequences

$$X = [A, B, C, D, B, A, C, D, F] \text{ and}$$
$$Y = [C, B, A, F, D, A, B]$$

### Solution

$$m = 9$$
$$n = 7$$

| $j$ | | 0 | 1 | 2 | 3 | 4 | 5 | 6 | 7 |
|---|---|---|---|---|---|---|---|---|---|
| $i$ | $y_i$ | | (C) | (B) | (A) | F | (D) | A | B |
| 0 | $x_i$ | 0 | 0 | 0 | 0 | 0 | 0 | 0 | 0 |
| 1 | A | 0 | ↑0 | ↑0 | ↖1 | ←1 | ←1 | ↖1 | ←1 |
| 2 | B | 0 | ↑0 | ↖1 | ↑1 | ↑1 | ↑1 | ↑1 | ↖2 |
| 3 | (C) | 0 | ↖1 | ↑1 | ↑1 | ↑1 | ↑1 | ↑1 | ↑2 |
| 4 | D | 0 | ↑1 | ↑1 | ↑1 | ↑1 | ↖2 | ←2 | ↑2 |
| 5 | (B) | 0 | ↑1 | ↖2 | ←2 | ←2 | ↑2 | ↑2 | ↖3 |
| 6 | (A) | 0 | ↑1 | ↑2 | ↖3 | ←3 | ←3 | ↖3 | ↑3 |
| 7 | C | 0 | ↖1 | ↑2 | ↑3 | ↑3 | ↑3 | ↑3 | ↑3 |
| 8 | (D) | 0 | ↑1 | ↑2 | ↑3 | ↑3 | ↖4 | ←4 | ←4 |
| 9 | F | 0 | ↑1 | ↑2 | ↑3 | ↖4 | ↑4 | ↑4 | ↑4 |

LCS = [C, B, A, D]

- The above figure shows the $c$ and $b$ tables computed by LCS-Length on the given sequences.
- The square in row $i$ and column $j$ contains the value of $c[i, j]$ and the appropriate arrow for the value of $b[i, j]$.
- The entry 4 in $c[9, 7]$ (the lower right hand corner of the table) is the length of an LCS ($CBAD$) of $X$ and $Y$.
- To reconstruct the elements of an LCS, follow the $b[i, j]$ arrows from the lower right hand corner. Each ↖ on the shaded sequence corresponds to an entry for which $x_i = y_j$ is a member of an LCS.

## 3.1.3 Knapsack Problem

- Given a set of items, each with a mass and a value.

- In knapsack problem, we have to determine the number of each item to include in a collection so that the total weight is less than or equal to a given limit and the total value is as large as possible.
- There are two versions of knapsack problem:
  - (i) 0 – 1 knapsack problem:
    In this, items are indivisible, i.e. we can either take one item or not.
  - (ii) Fractional knapsack problem:
    In this, items are divisible, i.e. we can take any fraction of item.
- Dynamic programming approach can provide solution to the knapsack problem:

**Step 1:** *Constructing an optimal substructure*

Consider a knapsack with maximum capacity $W$ and a set $S$ consisting of $n$ items. Each item $i$ has some weight $w_i$ and value $v_i$ (all $w_i$ and $W$ are integer values).

Let $w$ represents the maximum weight for each subset of items.

Now, the subproblem is to compute $s[k, w]$, i.e. to find an optimal solution for each item in a knapsack of size $w$.

**Step 2:** *A recursive solution*

Recursive formula for the sub-problem is

$$s[k, w] = \begin{cases} s[k-1, w] & \text{if } w_k > w \\ \max\{s[k-1, w], s[k-1, w-w_k]+v_k\} & \text{else} \end{cases}$$

**Step 3:** *Computing the optimal costs*

Following 0-1-knapsack algorithm is used to compute the optimal costs.

## Algorithm

```
0-1-knapsack (K, n, W)
Line 1      let r [1...n, 1...W] and s [0...n, 0...W] be new tables
Line 2      for i ← 1 to n
Line 3          s [i, 0] ← 0
Line 4      for w ← 0 to W
Line 5          s [0, w] ← 0
Line 6      for i ← 1 to n
Line 7          for w ← 1 to W
Line 8              if wᵢ ≤ w
```

Line 9                    if $v_i + s[i-1, w-w_i] > s[i-1, w]$
Line 10                       $s[i, w] \leftarrow v_i + s[i-1, w-w_i]$
Line 11                       $r[i, w] \leftarrow$ " $\nwarrow$ "
Line 12                    else $s[i, w] \leftarrow s[i-1, w]$
Line 13                       $r[i, w] \leftarrow$ "↑"
Line 14                else $s[i, w] \leftarrow s[i-1, w]$
Line 15                   $r[i, w] \leftarrow$ "↑"
Line 16        return $r$ and $s$

## Explanation

- The input to the 0-1-knapsack ($K, n, W$) algorithm is a knapsack $K$ with maximum capacity $W$ and consisting of $n$ items. Each item $i$ has some weight $w_i$ and value $v_i$.
- Line 1 indicates that the procedure maintains the table $r[1 \ldots n, 1 \ldots W]$ which helps in constructing an optimal solution. The procedure stores the $s[i, w]$ values in a table $s[0 \ldots n, 0 \ldots W]$, and it fills in the first row of $s$ from left to right and then the second row and so on.
- Line 2 indicates the beginning of for loop that ends with Line 3. This loop is applicable for $i$ equals to 1 to $n$.
- Line 3 indicates that $s[i, 0]$ is set to zero.
- Line 4 indicates the beginning of for loop that ends with Line 5. This loop is applicable for $w$ equals to 0 to $W$.
- Line 5 indicates that $s[0, w]$ is set to zero.
- Line 6 indicates the beginning of for loop that ends with Line 15. This loop is applicable for $i$ equals to 1 to $n$.
- Line 7 indicates the beginning of for loop that ends with Line 15. This loop is applicable for $w$ equals to 1 to $W$. Termination of this for loop causes the control to go back to Line 6.
- Line 8 checks the *if* condition. This condition is true when $w_i \leq w$. If this condition is true then the procedure checks the *if* condition of Line 9 else the control goes to Line 14 and the execution of Line 14 and Line 15 takes place.
- Line 9 checks the *if* condition. This condition is true if the sum of $v_i$ and the value at location $s[i-1, w-w_i]$ is greater than the value at location $s[i-1, w]$. If this condition is true then the execution of Line 10 and Line 11 takes place else the control goes to Line 12 and the execution of Line 12 and Line 13 takes place.
- Line 10 indicates that the location $s[i, w]$ is assigned a value which is equal to the sum of $v_i$ and the value at location $s[i-1, w-w_i]$.

- Line 11 indicates that ↖ sign is put at location $r[i, w]$.
- Line 12 indicates that the location $s[i, w]$ is assigned a value at location $s[i-1, w]$.
- Line 13 indicates that ↑ sign is put at location $r[i, w]$.
- Line 14 indicates that the location $s[i, w]$ is assigned a value at location $s[i-1, w]$.
- Line 15 indicates that ↑ sign is put at location $r[i, w]$.
- Line 16 returns tables $r$ and $s$.

### Analysis

- The running time of the procedure is $\Theta(nW)$ because each table entry takes $\Theta(1)$ time to compute.

*Step 4: Constructing an optimal solution from computed information*

The $r$ table returned by 0-1-knapsack algorithm enables us to construct an optimal solution.

We begin at $r[n, W]$ and trace through the table by following the arrows. Whenever, we encounter ↖ in entry $r[i, w]$, we include the corresponding item in the optimal collection of items.

### Example

Determine the items that should be included in the optimal knapsack:

| $i$ | $w_i$ | $v_i$ |
|:---:|:---:|:---:|
| 1 | 1 | 2 |
| 2 | 2 | 3 |
| 3 | 3 | 4 |
| 4 | 4 | 5 |
| 5 | 5 | 6 |

$$W = 6$$

## Solution

| $i/w$ | 0 | 1 | 2 | 3 | 4 | 5 | 6 |
|-------|---|---|---|---|---|---|---|
| 0 | 0 | 0 | 0 | 0 | 0 | 0 | 0 |
| (1) | 0 | ↖ 2 | ↖ 2 | ↖ 2 | ↖ 2 | ↖ 2 | ↖ 2 |
| (2) | 0 | ↑ 2 | 3 | 5 | 5 | 5 | 5 |
| (3) | 0 | ↑ 2 | ↑ 3 | ↑ 5 | 6 | 7 | 9 |
| 4 | 0 | ↑ 2 | ↑ 3 | ↑ 5 | ↑ 6 | ↑ 7 | ↑ 9 |
| 5 | 0 | ↑ 2 | ↑ 3 | ↑ 5 | ↑ 6 | ↑ 7 | ↑ 9 |

Optimal items = $\langle 3, 2, 1 \rangle$

Optimal value = $4 + 3 + 2 = 9$

- The above figure shows the $r$ and $s$ tables computed by 0-1-knapsack algorithm.
- The square in row $i$ and column $w$ contains the value of $s[i, w]$ and the appropriate arrow for the value of $r[i, w]$.
- The entry 9 in $s[5, 6]$ (the lower right hand corner of the table) is the optimal value.
- To include the optimal items, follow the $r[i, w]$ arrows from the lower right hand corner. Whenever we encounter ↖, we include the corresponding item.

## 3.2 GREEDY APPROACH

- In greedy approach one can make whatever choice seems best at the moment and then solve the subproblems that arise later.
- The choice made by greedy algorithm may depend on choices made so far but not on future choices.

- The greedy algorithm iteratively makes one greedy choice after other, reducing each given problem into a smaller one.
- In many problems a greedy strategy does not in general produce an optimal solution but nonetheless a greedy heuristic may yield locally optimal solutions, that approximate a global optimal solution, in reasonable time.
- A greedy approach can provide solutions to activity selection problem, construction of Huffman codes, single source shortest path problem, minimum spanning tree problem etc.

## 3.2.1 Activity Selection Problem

- Suppose, we have a set $S = \{a_1, a_2, ..., a_n\}$ of $n$ activities.
- Each activity $a_i$ has a start time $s_i$ and a finish time $f_i$, where $0 \leq s_i < f_i < \infty$.
- If selected, activity $a_i$ takes place during the half open time interval $[s_i, f_i)$.
- Activities $a_i$ and $a_j$ are compatible, if the intervals $[s_i, f_i)$ and $[s_j, f_j)$ do not overlap. That is $a_i$ and $a_j$ are compatible if $s_i \geq f_j$ or $s_j \geq f_i$.
- In the activity selection problem, we wish to select a maximum size subset of mutually compatible activities.
- We assume that the activities are sorted in monotonically increasing order of finish time.
- As $f_1 \leq f_2 \leq f_3 \cdots \leq f_{n-1} \leq f_n$.
- A greedy approach can provide solution to activity selection problem.

### Algorithm

Greedy-Activity-Selector (s, f)

Line 1      n ← length [s]
Line 2      A ← {a₁}
Line 3      k ← 1
Line 4      for m ← 2 to n
Line 5          if s [m] ≥ f [k]
Line 6              A ← A ∪ {a_m}
Line 7              k ← m
Line 8      return A

### Explanation

- The inputs to the Greedy-Activity-Selector (s, f) algorithm are an array s that stores the start times of the activities and an array f that stores the finish times of the activities.

- This algorithm assumes that the input activities are ordered by monotonically increasing finish time.
- Line 1 indicates that $n$ is equal to the length of array $s$.
- Line 2 indicates that set A is initialized by activity $a_1$, Set A is used to store the selected activities.
- Line 3 indicates that $k$ is initialized by the index of activity $a_1$, i.e. 1.
- Line 4 indicates the beginning of for loop that ends with Line 7. This loop is applicable for $m$ equals to 2 to $n$.
- Line 5 checks the *if* condition. If this condition is true then the execution of Line 6 and Line 7 takes place else the control goes back to Line 4. The *if* condition is true when the starting time $s[m]$ of activity $a_m$ is greater than or equal to the finishing time $f[k]$ of the activity most recently added to A.
- Line 6 indicates that activity $a_m$ is added to A.
- Line 7 indicates that $k$ is set to $m$.
- Line 8 returns set A.

## Analysis

- This procedure schedules a set of $n$ activities in $\Theta(n)$ time, assuming that the activities were already sorted initially by their finish times.

## Example

Illustrate the operation of Greedy-Activity-Selector $(s, f)$ algorithm on the following activities:

| Activities | Start time | Finish time |
|------------|------------|-------------|
| $a_1$ | 2 | 4 |
| $a_2$ | 1 | 5 |
| $a_3$ | 3 | 5 |
| $a_4$ | 4 | 6 |
| $a_5$ | 5 | 7 |
| $a_6$ | 6 | 8 |

## Solution

In the given problem the activities are arranged in increasing order of their finish times.

$$n = 6$$
$$A = \{a_1\}$$

$$k = 1$$

for $\quad\quad m = 2$

$$s[a_2] \not\geq f[a_1]$$

$$1 \not\geq 4$$

for $\quad\quad m = 3$

$$s[a_3] \not\geq f[a_1]$$

$$3 \not\geq 4$$

for $\quad\quad m = 4$

$$s[a_4] \geq f[a_1]$$

$$4 \geq 4$$

$$A = \{a_1, a_4\}$$

$$k = 4$$

for $\quad\quad m = 5$

$$s[a_5] \not\geq f[a_4]$$

$$5 \not\geq 6$$

for $\quad\quad m = 6$

$$s[a_6] \geq f[a_4]$$

$$6 \geq 6$$

$$A = \{a_1, a_4, a_6\}$$

$$k = 6$$

The resultant set of selected activities is

$\{a_1, a_4, a_6\}$

## 3.2.2 Huffman Codes

- The Huffman codes are the 'prefix codes' that are optimum for a given set of probabilities.
- Huffman procedure is based on two observations regarding optimum prefix codes.

  (i) In an optimum code, symbols that occur more frequently (have a higher probability of occurrence), will have shorter code words than symbols that occur less frequently.

  (ii) In an optimum code, the two symbols that occur least frequently will have the same length.

- A 'prefix code' is that code in which no code-word is a 'prefix' to another code-word.

- Suppose, we have two binary code-words $a$ and $b$, where $a$ is $m$ bits long, $b$ is $n$ bits long and $m < n$. If the first $m$ bits of $b$ are identical to $a$, then $a$ is called a prefix of $b$.

e.g. Let $\quad\quad\quad\quad a = 01011$
$$b = 0101100$$

then $a$ is a prefix of $b$.

## Algorithm

Huffman (C)

| | |
|---|---|
| Line 1 | $n \leftarrow |C|$ |
| Line 2 | $Q \leftarrow C$ |
| Line 3 | for $i \leftarrow 1$ to $n - 1$ |
| Line 4 | allocate a new node z |
| Line 5 | left [z] $\leftarrow$ x $\leftarrow$ Extract-Min (Q) |
| Line 6 | right [z] $\leftarrow$ y $\leftarrow$ Extract-Min (Q) |
| Line 7 | freq [z] $\leftarrow$ freq [x] + freq [y] |
| Line 8 | Insert (Q, z) |
| Line 9 | return Extract-Min (Q) |

## Explanation

- The Huffman (C) procedure generates an optimal prefix code called as Huffman code.
- The input to the Huffman (C) procedure is a set C of $n$ characters.
- Each character $c \in C$ is an object which is associated with an attribute freq [c] that gives its frequency.
- The above algorithm constructs a tree T corresponding to the optimal code in a bottom up fashion.
- Line 1 indicates that $n$ stores the number of characters in set C.
- Line 2 indicates that the priority queue Q is initialized with the characters in C.
- Line 3 indicates the beginning of for loop that ends with Line 8. This for loop is applicable for $i$ equals to 1 to $n - 1$.
- Line 4 allocates a new node z.
- Line 5 indicates that x becomes the left child of node z. x is the element of Q having smallest key value, i.e. lowest frequency. x is removed and returned by the Extract-Min (Q) operation.
- Line 6 indicates that y becomes the right child of node z. y is the element of Q having smallest key value, i.e. lowest frequency. y is removed and returned by the Extract-Min (Q) operation.
- Line 7 indicates that the frequency of z is equals to the sum of frequencies of x and y.
- Line 8 inserts z in priority queue Q.

- Line 9 after $n - 1$ merges, returns the one node left in the queue, which is the root of the code tree.

## Analysis

- The queue $Q$ in Line 2 is initialized in $O(n)$ time using the Build-Min-Heap procedure.
- The for loop in Lines 3-8 executes exactly $n - 1$ times, and since each heap operation requires time $O(\lg n)$, the loop contributes $O(n \lg n)$ to the running time.
- Thus, the total running time of Huffman algorithm on a set of $n$ characters is $O(n \lg n)$.

## Example

Generate an optimal Huffman code for the following set of frequencies:

$a$: 4, $b$: 12, $c$: 8, $d$: 5, $e$: 14.

## Solution

Huffman (C)

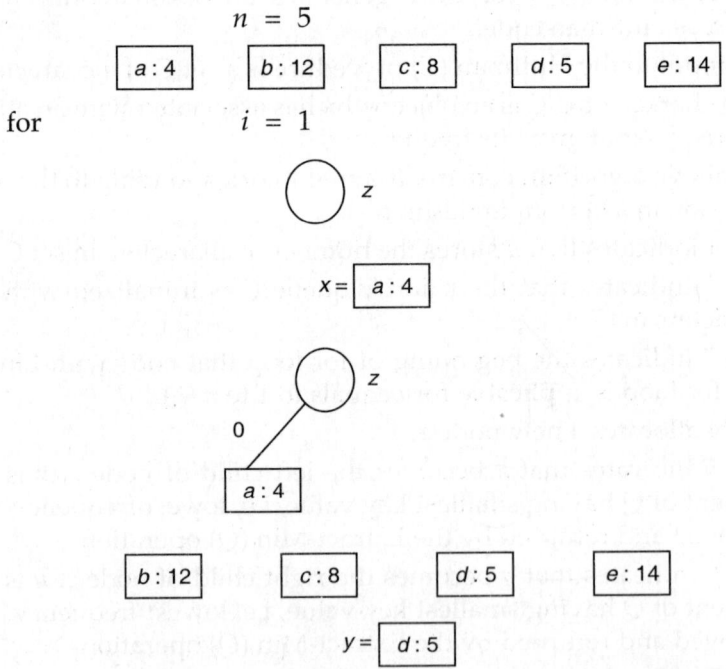

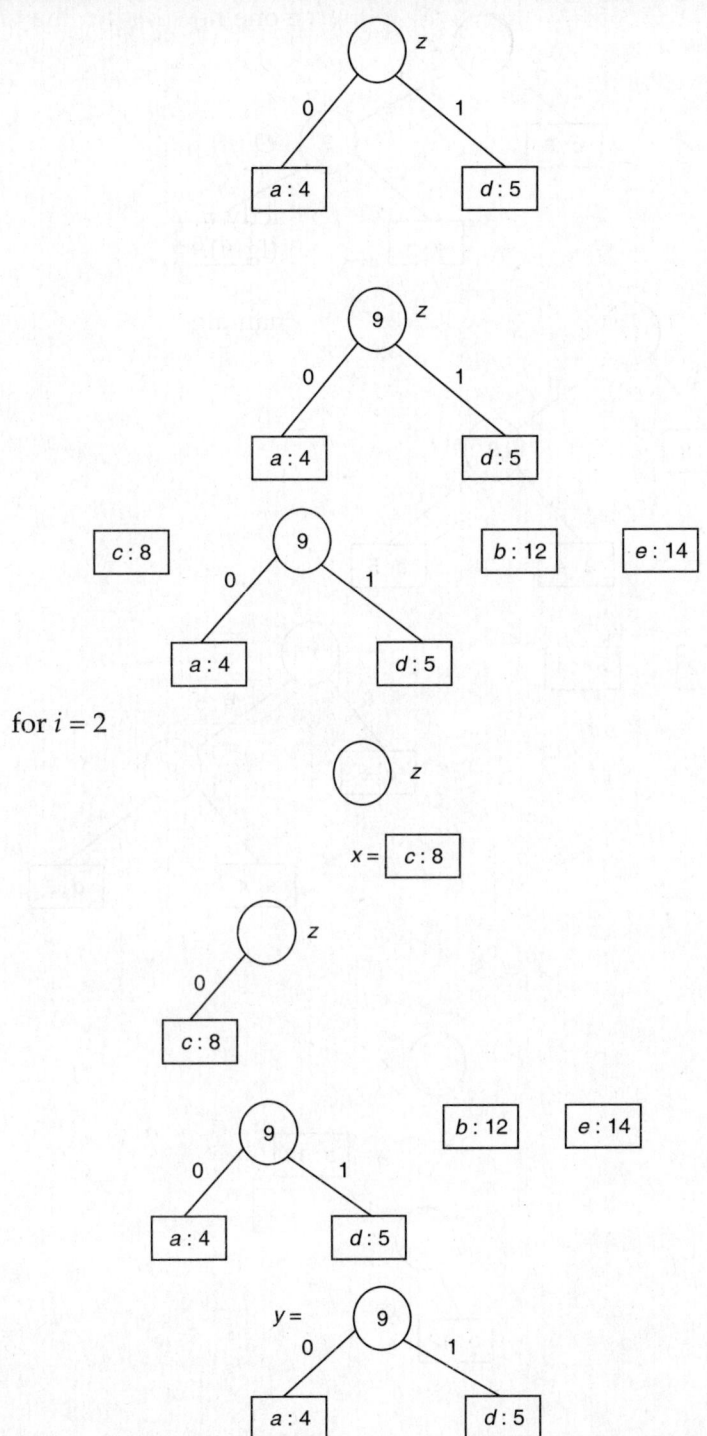

for $i = 2$

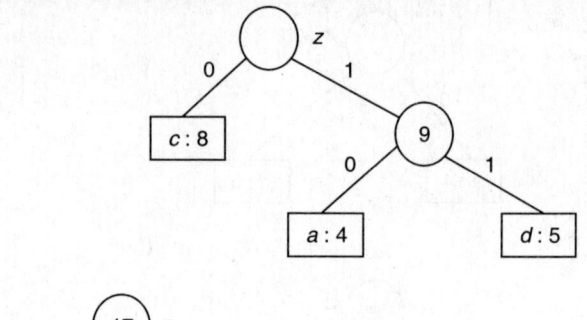

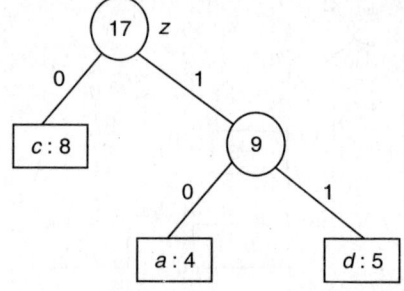

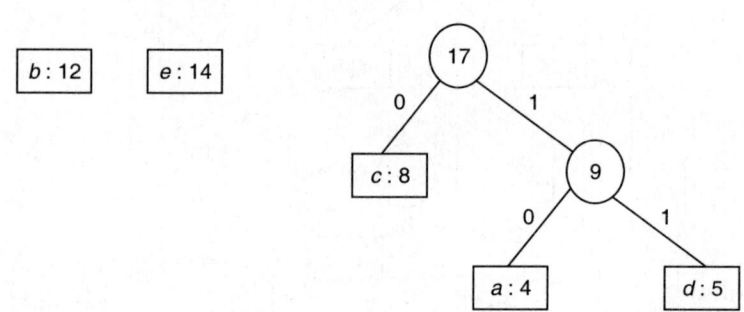

for                          $i = 3$

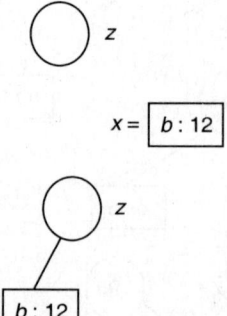

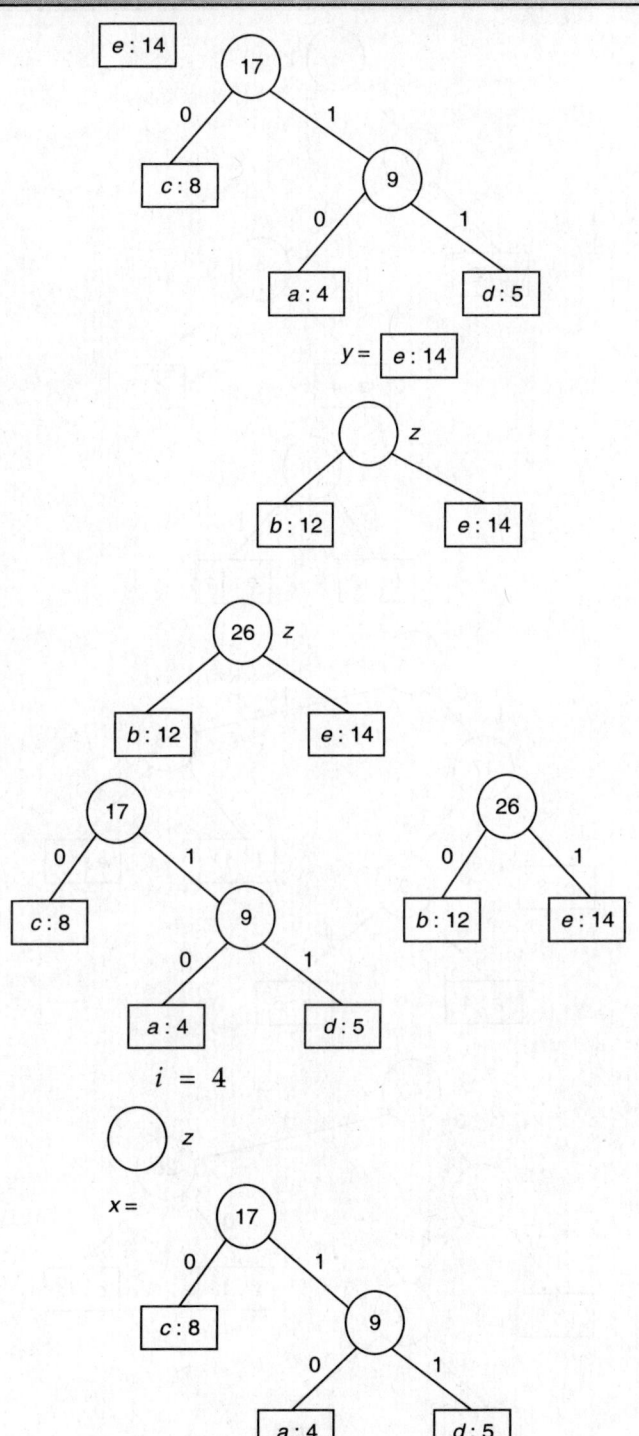

$$y = \boxed{e : 14}$$

for          $i = 4$

$$x =$$

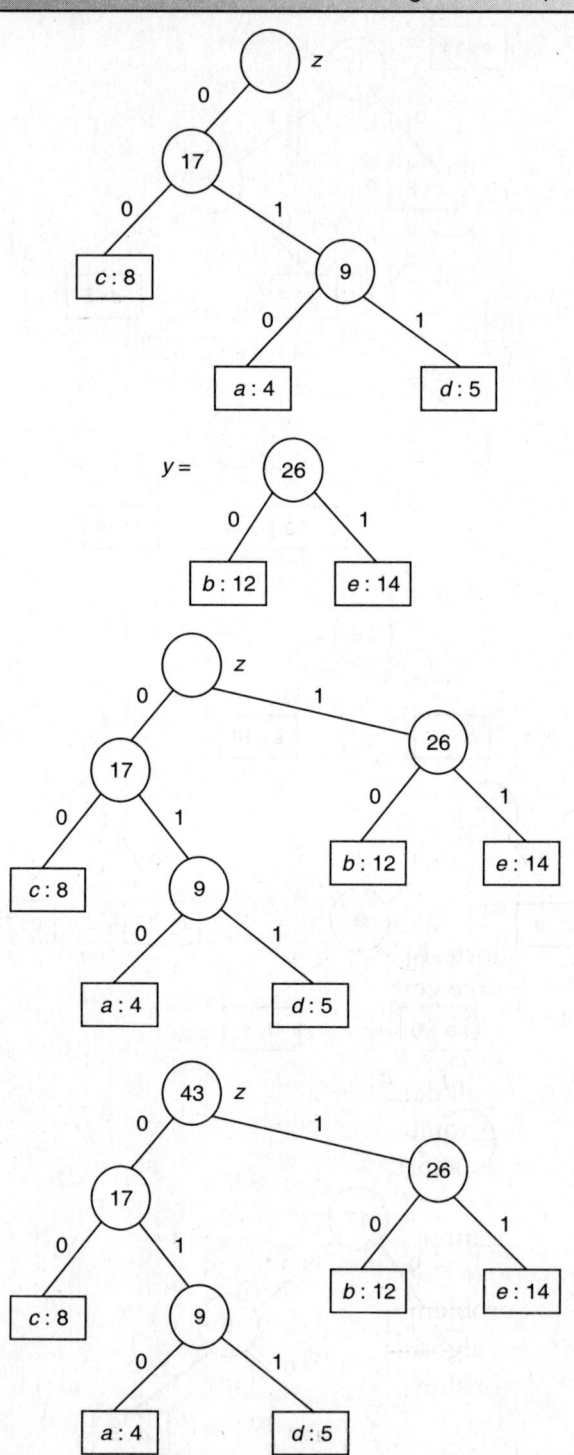

After the termination of for loop:

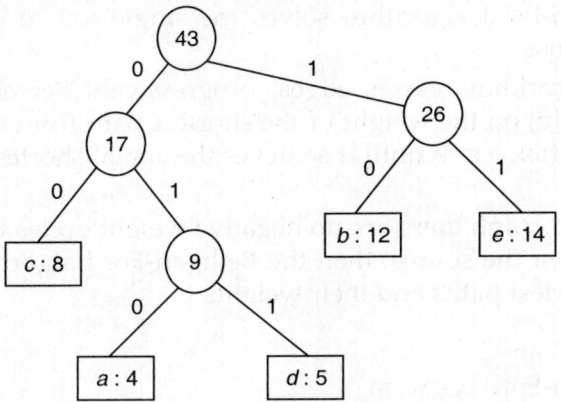

The tree corresponding to the optimal prefix code $a = 010$, $b = 10$, $c = 00$, $d = 011$, $e = 11$.

### 3.2.3 Single-source Shortest Paths

- A shortest path from vertex $u$ to vertex $v$ is defined as any path $p$ with weight $w(p) = \delta(u, v)$

  where, $\delta(u, v)$ is known as the shortest path weight

  $$\delta(u, v) = \begin{cases} \min\{w(p) : u \xrightarrow{\ p\ } v\} & \text{if there is path from } u \text{ to } v \\ \infty & \text{otherwise} \end{cases}$$

  **Problem**

  Consider a graph $G = (V, E)$

  Single-source shortest paths problem asks to find a shortest path from a given source vertex $s \in V$ to each vertex $v \in V$.

- If the graph $G = (V, E)$ contains no negative weight cycles reachable from the source $s$, then for all $v \in V$, the shortest path weight $\delta(s, v)$ remains well defined, even if it has a negative value.

  But, if a graph contains a negative weight cycle reachable from source $s$, then the shortest path weights are not well defined. In this case $\delta(s, v) = -\infty$

- A shortest path cannot contain a cycle.

- Bellman-Ford and Dijkstra's algorithms solve the single source shortest paths problem.

- In Bellman-Ford algorithm, edge weight can be negative.

- In Dijkstra's algorithm, edge weight should be non-negative.

### 3.2.3.1 Bellman-Ford algorithm

- Bellman-Ford algorithm solves the single-source shortest paths problems.
- The algorithm relaxes edges, progressively decreasing an estimate $d[v]$ on the weight of the shortest path from the source $s$ to each vertex $v \in V$ until it achieves the actual shortest path weight $\delta(s, v)$.
- If in the graph there are no negative weight cycles that are reachable from the source, then the Bellman-Ford algorithm produces the shortest paths and their weights.

### Algorithm

Bellman-Ford (G, w, s)

| | |
|---|---|
| Line 1 | Initialize-Single-Source (G, s) |
| Line 2 | for $i \leftarrow 1$ to $\|V[G]\| - 1$ |
| Line 3 | for each edge $(u, v) \in E[G]$ |
| Line 4 | Relax $(u, v, w)$ |
| Line 5 | for each edge $(u, v) \in E[G]$ |
| Line 6 | if $d[v] > d[u] + w(u, v)$ |
| Line 7 | return false |
| Line 8 | return true |

### Explanation

- The inputs to the Bellman-Ford $(G, w, s)$ algorithm are a directed graph $G$ having weighted edges and a source vertex $s$. $w$ denotes the weight of an edge. Edge weights may be negative.
- Line 1 calls the procedure Initialize-Single-Source $(G, s)$.
- Line 2 indicates the beginning of for loop that ends with Line 4. This loop is applicable for $i$ equals to 1 to $|V[G]| - 1$. $|V[G]|$ indicates the number of vertices in the graph $G$.
- Line 3 indicates the beginning of for loop that ends with Line 4. This loop is applicable for each edge $(u, v)$ that belongs to the set of edges of graph $G$. Termination of this for loop causes the control to go back to Line 2.
- Line 4 calls the procedure Relax $(u, v, w)$.
- Line 5 indicates the beginning of for loop that ends with Line 7. This loop is applicable for each edge $(u, v)$ that belongs to the set of edges of graph $G$.
- Line 6 checks the *if* condition. This condition is true if $d[v]$ is greater than the sum of $d[u]$ and $w(u, v)$.

Attribute $d\,[u]$ holds the shortest path weight from source $s$ to vertex $u$.

Attribute $d\,[v]$ is an upper bound on the weight of the shortest path from source s to $v$.

$w\,(u, v)$ denotes the weight of edge $(u, v)$. If this condition is true then the execution of Line 7 takes place else the control goes back to Line 5.

- Line 7 returns the boolean value False. It means that the graph $G$ contains negative weight cycles reachable from source $s$.
- Line 8 returns the boolean value True. It means that the graph $G$ contains no negative weight cycle reachable from source $s$. The execution of Line 8 takes place only when for every edge the *if* condition of Line 6 fails.

## Analysis

This algorithm runs in $O\,(|V|\;|E|)$ time because the initialization in Line 1 takes $\Theta\,(|V|)$ time, each of the $|V| - 1$ passes over the edges in Lines 2-4 takes $\Theta\,(|E|)$ time and the for loop of Lines 5-7 takes $O\,(|E|)$ time.

## Algorithm

        Initialize-Single-Source (G, s)
    Line 1        for each vertex v ∈ V [G]
    Line 2              d [v] ← ∞
    Line 3              π [v] ← nil
    Line 4        d [s] ← 0

## Explanation

- Line 1 indicates the beginning of for loop that ends with Line 3. This loop is applicable for each vertex $v$ that belongs to the set of vertices of graph G.
- Line 2 indicates that $d[v]$ is set to ∞. It means that there is no path from s to $v$.
- Line 3 indicates that $\pi[v]$ is set to nil. Attribute $\pi[v]$ stores the parent of vertex $v$.
- Line 4 indicates that $d[s]$ is set to zero. It means that the shortest path weight from source s to itself is zero.

## Analysis

- This algorithm takes $\Theta\,(|V|)$ time.

## Algorithm

Relax (u, v, w)

Line 1      if $d[v] > d[u] + w$ (u, v)
Line 2          $d[v] \leftarrow d[u] + w$ (u, v)
Line 3          $\pi[v] \leftarrow u$

## Explanation

- Line 1 checks the *if* condition. This condition is true if $d[v]$ is greater than the sum of $d[u]$ and $w$ (u, v).

  The *if* condition tests whether we can improve the shortest path to $v$ found so far by going through $u$.

  If this condition is true then the execution of Line 2 and Line 3 takes place.

- Line 2 updates the value of $d[v]$ by assigning it a new value which is equal to the sum of $d[u]$ and $w$ (u, v).

- Line 3 indicates that vertex $u$ becomes the parent of vertex $v$.

- The process of relaxing an edge is used to estimate the shortest path.

## Analysis

- The above procedure performs a relaxation step on edge (u, v) in $O(1)$ time.

## Example

Illustrate the operation of Bellman-Ford algorithm on the following directed graph.

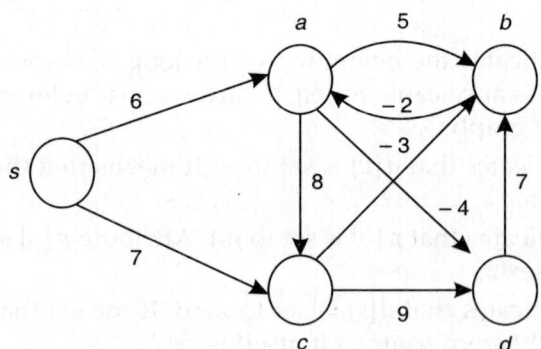

## Solution

Bellman-Ford (G, w, s)
Initialize-Single-Source (G, s)

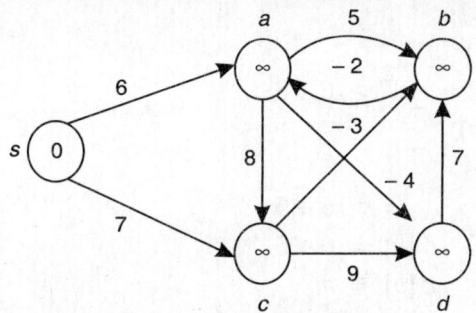

The *d* values appear within the vertices

The first for loop of Bellman-Ford algorithm runs for *i* equals to 1 to 4

for                          $i = 1$

for edge $(s, a)$

Relax $(s, a, 6)$

$$\infty > 0 + 6$$
$$d[a] = 0 + 6$$
$$\pi[a] = s$$

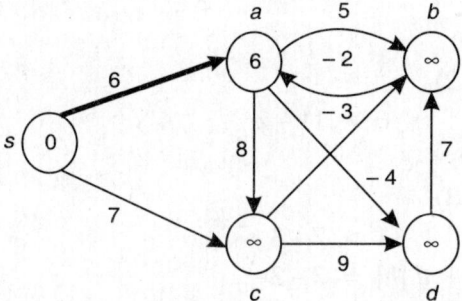

for edge $(s, c)$

Relax $(s, c, 7)$

$$\infty > 0 + 7$$
$$d[c] = 0 + 7$$
$$\pi[c] = s$$

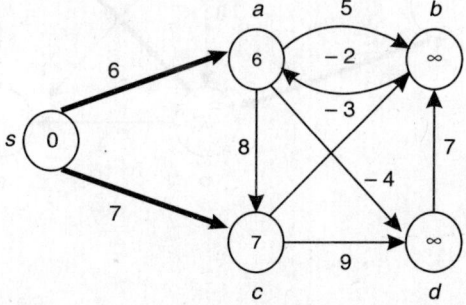

for edge $(a, c)$
Relax $(a, c, 8)$

$$7 \not> 6 + 8$$

for edge $(a, b)$
Relax $(a, b, 5)$

$$\infty > 6 + 5$$
$$d[b] = 6 + 5$$
$$\pi[b] = a$$

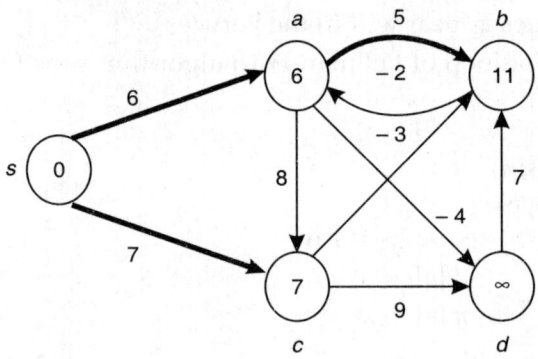

for edge $(b, a)$
Relax $(b, a, -2)$

$$6 \not> 11 - 2$$

for edge $(c, b)$
Relax $(c, b, -3)$

$$11 > 7 - 3$$
$$d[b] = 7 - 3$$
$$\pi[b] = c$$

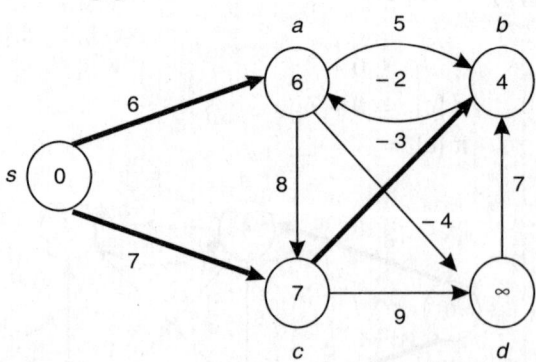

for edge $(a, d)$
Relax $(a, d, -4)$

$$\infty > 6 - 4$$
$$d\,[d] = 6 - 4$$
$$\pi\,[d] = a$$

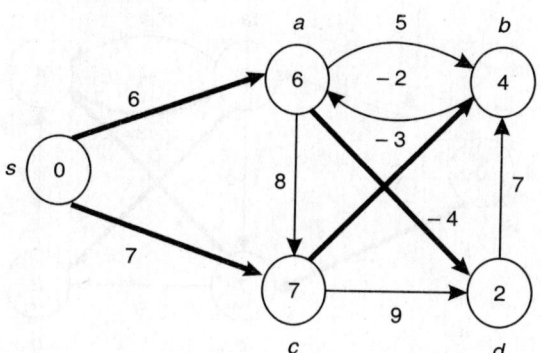

for edge $(c, d)$
Relax $(c, d, 9)$

$$2 \ngtr 7 + 9$$

for edge $(d, b)$
Relax $(d, b, 7)$

$$4 \ngtr 2 + 7$$

for                $i = 2$
for edge $(s, a)$
Relax $(s, a, 6)$

$$6 \ngtr 0 + 6$$

for edge $(s, c)$
Relax $(s, c, 7)$

$$7 \ngtr 0 + 7$$

for edge $(a, b)$
Relax $(a, b, 5)$

$$4 \ngtr 6 + 5$$

for edge $(a, c)$
Relax $(a, c, 8)$

$$7 \ngtr 6 + 8$$

for edge $(a, d)$
Relax $(a, d, -4)$

$$2 \ngtr 6 - 4$$

for edge $(b, a)$
Relax $(b, a, -2)$

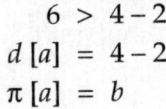

$$6 > 4 - 2$$
$$d[a] = 4 - 2$$
$$\pi[a] = b$$

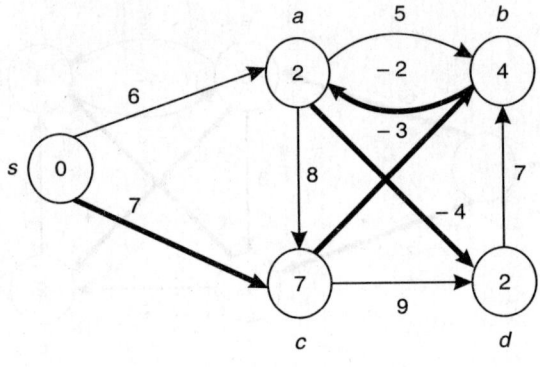

for edge $(c, b)$
Relax $(c, b, -3)$

$$4 \ngtr 7 - 3$$

for edge $(c, d)$
Relax $(c, d, 9)$

$$2 \ngtr 7 + 9$$

for edge $(d, b)$
Relax $(d, b, 7)$

$$4 \ngtr 2 + 7$$

for        $i = 3$

for edge $(s, a)$
Relax $(s, a, 6)$

$$2 \ngtr 0 + 6$$

for edge $(s, c)$
Relax $(s, c, 7)$

$$7 \ngtr 0 + 7$$

for edge $(a, b)$
Relax $(a, b, 5)$

$$4 \ngtr 2 + 5$$

for edge $(a, c)$
Relax $(a, c, 8)$

$$7 \ngtr 2 + 8$$

for edge $(a, d)$
Relax $(a, d, -4)$

$$2 > 2-4$$
$$d[d] = 2-4$$
$$\pi[d] = a$$

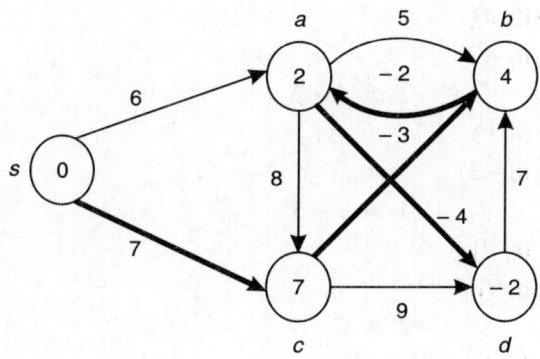

for edge $(b, a)$
Relax $(b, a, -2)$

$$2 \not> 4-2$$

for edge $(c, b)$
Relax $(c, b, -3)$

$$4 \not> 7-3$$

for edge $(c, d)$
Relax $(c, d, 9)$

$$-2 \not> 7+9$$

for edge $(d, b)$
Relax $(d, b, 7)$

$$4 \not> -2+7$$

for          $i = 4$

for edge $(s, a)$
Relax $(s, a, 6)$

$$2 \not> 0+6$$

for edge $(s, c)$
Relax $(s, c, 7)$

$$7 \not> 0+7$$

for edge $(a, b)$
Relax $(a, b, 5)$

$$4 \not> 2+5$$

for edge $(a, c)$
Relax $(a, c, 8)$

$$7 \not> 2+8$$

for edge $(a, d)$

Relax $(a, d, -4)$

$$-2 \ngtr 2 - 4$$

for edge $(b, a)$

Relax $(b, a, -2)$

$$2 \ngtr 4 - 2$$

for edge $(c, b)$

Relax $(c, b, -3)$

$$4 \ngtr 7 - 3$$

for edge $(c, d)$

Relax $(c, d, 9)$

$$-2 \ngtr 7 + 9$$

for edge $(d, b)$

Relax $(d, b, 7)$

$$4 \ngtr -2 + 7$$

Now, we consider the execution of for loop of Line 5

for edge $(s, a)$

$$2 \ngtr 0 + 6$$

for edge $(s, c)$

$$7 \ngtr 0 + 7$$

for edge $(a, b)$

$$4 \ngtr 2 + 5$$

for edge $(a, c)$

$$7 \ngtr 2 + 8$$

for edge $(a, d)$

$$-2 \ngtr 2 - 4$$

for edge $(b, a)$

$$2 \ngtr 4 - 2$$

for edge $(c, b)$

$$4 \ngtr 7 - 3$$

for edge $(c, d)$

$$-2 \ngtr 7 + 9$$

for edge $(d, b)$

$$4 \ngtr -2 + 7$$

After execution of Line 8, the Bellman-Ford algorithm returns a Boolean value True for this example.

## 3.2.3.2 Dijkstra's algorithm

- Dijkstra's algorithm solves the single-source shortest path problems.
- Dijkstra's algorithm maintains a set $S$ of vertices. The algorithm repeatedly selects the vertex $u \in V - S$ with the minimum shortest path estimate, adds $u$ to $S$ and relaxes all edges leaving $u$.
- Dijkstra's algorithm uses greedy approach.

**Algorithm**

      Dijkstra (G, w, s)

Line 1        Initialize-Single-Source (G, s)

Line 2        $S \leftarrow \phi$

Line 3        $Q \leftarrow V [G]$

Line 4        while $Q \neq \phi$

Line 5           $u \leftarrow$ Extract-Min (Q)

Line 6           $S \leftarrow S \cup \{u\}$

Line 7           for each vertex v $\in$ Adj [u]

Line 8               Relax (u, v, w)

**Explanation**

- The inputs to Dijkstra's algorithm are a directed weighted graph $G$ and a source vertex $s$. $w$ denotes the weights of edges. All weights of edges should be non-negative.
- Line 1 calls the procedure Initialize-Single-Source $(G, s)$.
- Line 2 indicates that $S$ is set to nil. $S$ is a set of vertices whose final shortest path weights from source $s$ have already been determined.
- Line 3 initializes the priority queue $Q$ by filling it with all the vertices of graph $G$.
- Line 4 indicates the beginning of while loop that ends with Line 8. This loop is applicable till the priority queue becomes empty.
- Line 5 indicates that $u$ is the vertex extracted by the Extract-Min $(Q)$ operation. The Extract-Min $(Q)$ operation removes and returns the element of $Q$ with the smallest key.
- The Extract-Min $(Q)$ operation extracts $u$ from $Q = V - S$.
- Line 6 indicates that the vertex $u$ extracted by Extract-Min $(Q)$ operation is added to set $S$.
- Line 7 indicates the beginning of for loop that ends with Line 8. This loop is applicable for each vertex $v$ that belongs to the adjacency list of vertex $u$.
- Line 8 calls the procedure Relax $(u, v, w)$.

## Analysis

The Dijkstra's algorithm maintains the min-priority queue, $Q$ by calling three priority queue operations:

   (i)  Insert (implicit in Line 3)

  (ii)  Extract-Min (Line 5)

 (iii)  Decrease-key (implicit in Relax procedure called in Line 8).

- The algorithm calls both Insert and Extract-Min once per vertex.
  Because each vertex $u \in V$ is added to set $S$ exactly once, each edge in the adjacency list Adj $[u]$ is examined in the for loop of Lines 7-8 exactly once during the course of algorithm.

- Since the total number of edges in all the adjacency lists is $|E|$, this for loop iterates a total of $|E|$ times and thus, the algorithm calls Decrease-key at most $|E|$ times overall.

- The running time of Dijkstra's algorithm depends on how we implement the min-priority queue.

- If we implement $Q$ as a binary min heap, then each Extract-Min operation takes $O(\lg |V|)$ time.
  The time to build the min-heap is $O(|V|)$.
  Each Decrease-key operation takes $O(\lg |V|)$ time and there are still at most $|E|$ such operations.
  Therefore, the total running time is $O((|V| + |E|) \lg |V|)$, which is
  $O(|E| \lg |V|)$, if all vertices are reachable from the source.

## Example

Show the result of Dijkstra's algorithm on the following graph:

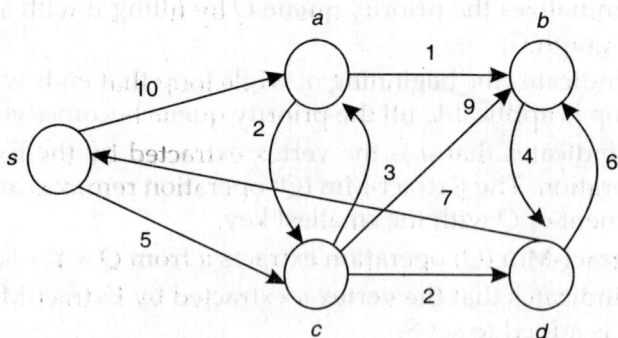

## Solution

    Initialize-Single-Source $(G, s)$

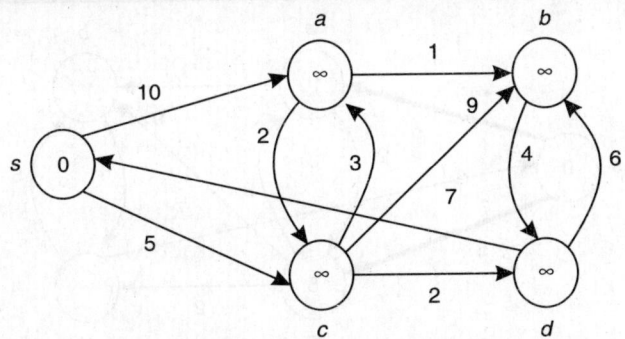

The $d$ values appear within the vertices. These $d$ values act as keys
for the vertices.

$$S = \phi$$

| $u$ | $s$ | $a$ | $b$ | $c$ | $d$ |
|---|---|---|---|---|---|
| key $[u]$ | 0 | $\infty$ | $\infty$ | $\infty$ | $\infty$ |

$$Q \neq \phi$$
$$u = s \ [\text{Extract-Min}(Q) \text{ operation extracts the}$$
$$\text{element } s \text{ having minimum key value zero}]$$
$$S = \{s\}$$

Vertices adjacent to s are $a$ and $c$

for                    $v = a$

Relax $(s, a, 10)$

$$\infty > 0 + 10$$
$$d\,[a] = 0 + 10$$
$$\pi\,[a] = s$$

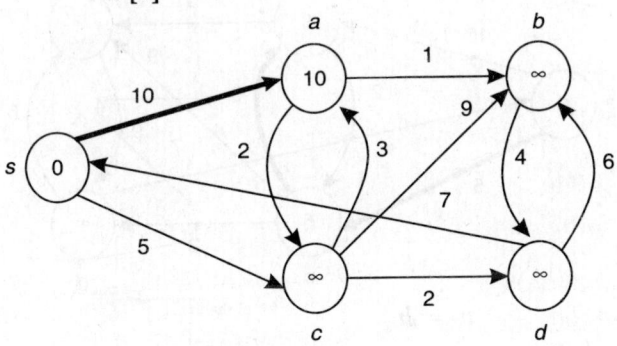

for                    $v = c$

Relax $(s, c, 5)$

$$\infty > 0 + 5$$
$$d\,[c] = 0 + 5$$
$$\pi\,[c] = s$$

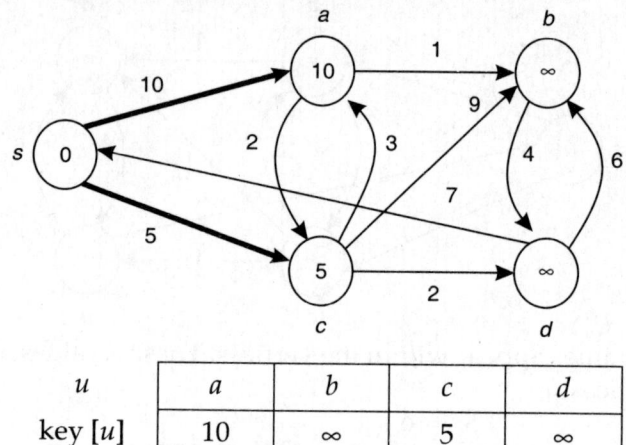

| $u$ | $a$ | $b$ | $c$ | $d$ |
|---|---|---|---|---|
| key [$u$] | 10 | ∞ | 5 | ∞ |

$$Q \neq \phi$$

$u = c$ [Extract-Min ($Q$) operation extracts the element $c$ having minimum key value 5]

$$S = \{s, c\}$$

Vertices adjacent to $c$ are $a$, $b$ and $d$.

for                    $v = a$

Relax ($c, a, 3$)

$$10 > 5 + 3$$
$$d[a] = 8$$
$$\pi[a] = c$$

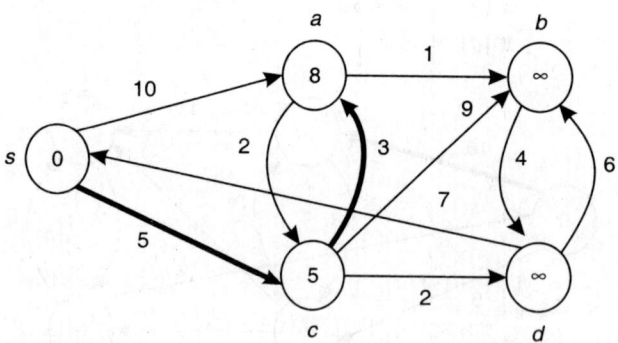

for                    $v = b$

Relax ($c, b, 9$)

$$\infty > 5 + 9$$
$$d[b] = 13$$
$$\pi[b] = c$$

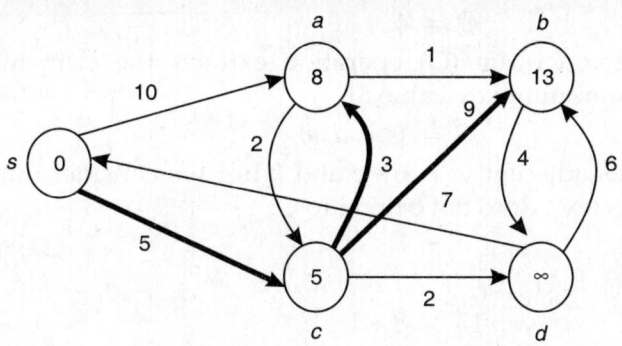

for                    $v = d$

Relax $(c, d, 2)$

$$\infty > 5 + 2$$
$$d[d] = 5 + 2$$
$$\pi[d] = c$$

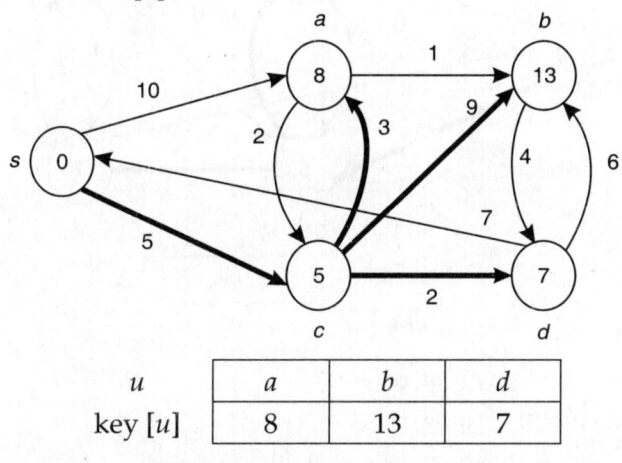

| $u$ | $a$ | $b$ | $d$ |
|-----|-----|-----|-----|
| key $[u]$ | 8 | 13 | 7 |

$$Q \neq \phi$$

$u = d$ [Extract-Min $(Q)$ operation extracts the element $d$ having minimum key value 7]

$$S = \{s, c, d\}$$

Vertices adjacent to $d$ are $s$ and $b$ but we consider only vertex $b$ because vertex $s$ does not belong to $Q$.

for                    $v = b$

Relax $(d, b, 6)$

$$13 \not> 7 + 6$$

| $u$ | $a$ | $b$ |
|-----|-----|-----|
| key$[u]$ | 8 | 13 |

$$Q \neq \phi$$

$u = a$ [Extract-Min $(Q)$ operation extracts the element $a$ having minimum key value 8]

$$S = \{s, c, d, a\}$$

Vertices adjacent to $a$ are $c$ and $b$ but we consider only vertex $b$ because vertex $c$ does not belong to $Q$.

for $\qquad v = b$

Relax $(a, b, 1)$

$$13 > 8 + 1$$
$$d[b] = 9$$
$$\pi[b] = a$$

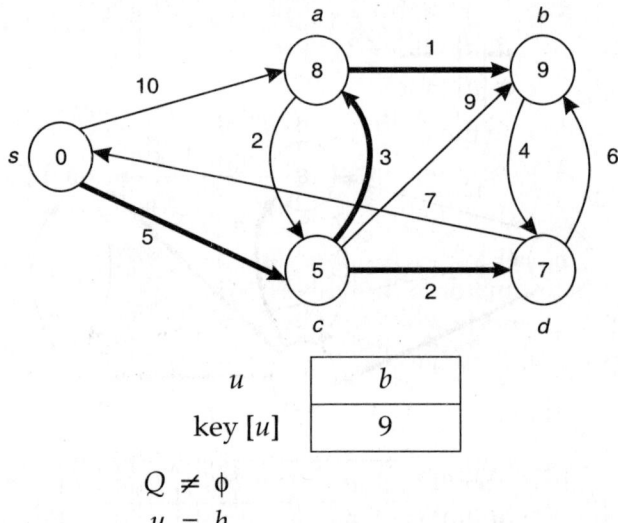

| $u$ | $b$ |
|---|---|
| key $[u]$ | 9 |

$$Q \neq \phi$$
$$u = b$$
$$S = \{s, c, d, a, b\}$$

Vertex adjacent to $b$ is $d$, but we don't consider it because it does not belong to $Q$.

Now $\qquad\qquad Q == \phi$

So, while loop terminates.

## 3.2.4 Minimum Spanning Trees

Consider a connected, undirected graph $G = (V, E)$ having weighted edges.

$V$ denotes vertices of graph $G$

$E$ denotes edges of graph $G$

$w(u, v)$ denotes the weight of an edge $(u, v)$

A minimum spanning tree is a tree that connects all the vertices and having minimum weight.

A minimum spanning tree may comprise of some or all of the edges of a graph.

**Example**

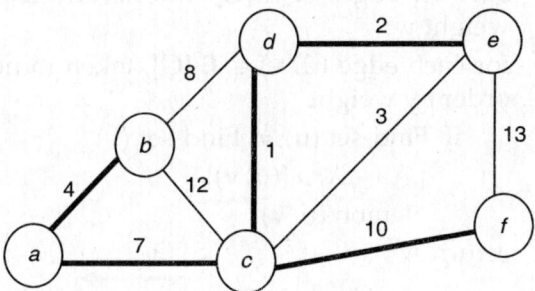

Dark lines denote the minimum spanning tree with weight 24.

*Generic method for growing a minimum spanning tree:*

- The generic method for growing a minimum spanning tree manages a set of edges $A$.
- As the algorithm for minimum spanning tree proceeds, a 'safe edge' keeps on adding to $A$.
- Kruskal's algorithm and Prim's algorithm are the minimum spanning tree algorithms.
- In Kruskal's algorithm, the set $A$ is a forest whose vertices are all those of the given graph.
- In Prim's algorithm, the set $A$ forms a single tree.
- Both these algorithms use greedy approach.

### 3.2.4.1 Kruskal's algorithm

- Kruskal's algorithm is a minimum spanning tree algorithm.
- The Kruskal's algorithm finds a safe edge and adds it to the growing forest.
- Safe edge is an edge of least weight that connects two distinct components.
- The algorithm inspects all the edges that connect any two trees in the forest and then out of them selects a safe edge.

## Algorithm

MST-Kruskal (G, w)

Line 1　　A ← φ

Line 2　　for each vertex v ∈ V [G]

Line 3　　　　Make-set (v)

Line 4　　Sort the edges of E [G] into non-decreasing order by weight w.

Line 5　　for each edge (u, v) ∈ E [G], taken in non-decreasing order by weight

Line 6　　　　if Find-set (u) ≠ Find-set (v)

Line 7　　　　　　A ← A ∪ {(u, v)}

Line 8　　　　　　Union (u, v)

Line 9　　return A

## Explanation

- The input to the above algorithm is a graph $G$ having weighted edges. $w$ denotes weight.
- Line 1 indicates that the date structure $A$ is set to an empty set. $A$ is used to store the disjoint set of elements.
- Line 2 indicates the beginning of for loop that ends with Line 3. This loop is applicable for every vertex $v$ that belongs to the set of vertices of graph $G$.
- Line 3 calls the operation Make-set ($v$). This operation creates a new set having only a single member $v$.
- Line 4 sorts the edges of the graph in increasing order of their weights.
- Line 5 indicates the beginning of for loop that ends with Line 8. This loop is applicable for each edge of the graph $G$ taken in increasing order by weight.
- Line 6 checks the *if* condition. If this condition is true then the execution of Line 6 and Line 7 takes place else the control goes back to Line 5. The *if* condition is true when Find-set ($u$) ≠ Find-set ($v$)

  In general Find-set (x) returns a pointer to the represerive of the set containing x.

  Find-set ($u$) ≠ Find-set ($v$), it means that the two vertices $u$ and $v$ do not belong to the same tree.
- Line 7 adds the edge ($u, v$) to set $A$.
- Line 8 calls the operation Union ($u, v$). This operation unites the sets containing $u$ and $v$ into a new set which is the union of these two sets.
- Line 9 returns $A$.

## Analysis

- Time to initialize the set $A$ in Line 1 is $O(1)$.
- Time to sort the edges in Line 4 is $O(|E| \lg |E|)$.
- The for loop of Lines 5-8 performs $O(|E|)$ Find-set and Union operations on the disjoint-set forest. Along with the $|V|$ Make-set operations, these take a total of $O((|V| + |E|) \alpha(|V|))$ time, where $\alpha$ is the very slowly growing function.
- Since $G$ is connected, $|E| \geq |V| - 1$, so the disjoint set operations take $O(|E| \alpha(|V|))$ time.
- Moreover, since $\alpha(|V|) = O(\lg |V|) = O(\lg |E|)$, the total running time of Kruskal's algorithm is $O(|E|) \lg |E|$.
- Observing that $|E| < |V|^2$, we have $\lg |E| = O(\lg |V|)$, and so we can restate the running time of Kruskal's algorithm as $O(|E| \lg |V|)$.

## Example

Show the result of Kruskal's algorithm on the following graph.

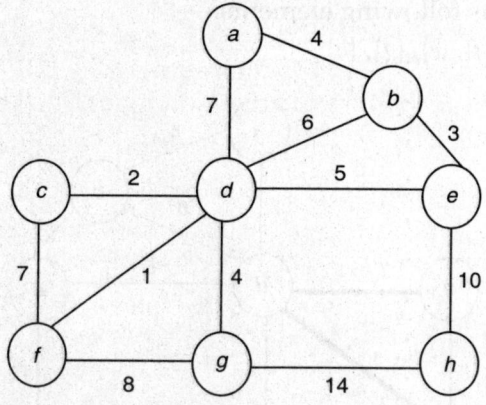

## Solution

MST-Kruskal $(G, w)$

After the execution of first for loop, $A$ has following elements:

$\{a\}, \{b\}, \{c\}, \{d\}, \{e\}, \{f\}, \{g\}, \{h\}$

After the execution of Line 4

$(d, f), (c, d), (b, e), (a, b), (d, g), (d, e), (b, d), (a, d), (c, f), (f, g), (e, h), (g, h)$

Consider edge $(d, f)$

$d$ and $f$ do not belong to the same tree, so this edge can be added to A.

Now, A has following elements:

{a}, {b}, {c}, {d, f}, {e}, {g}, {h}

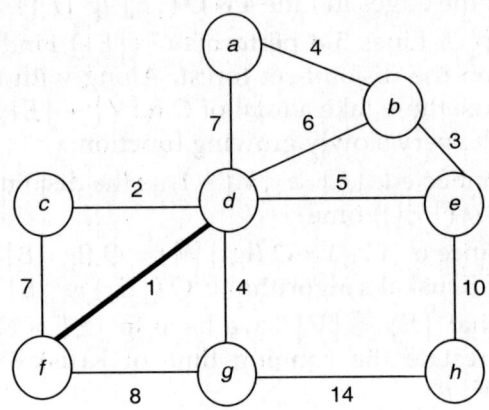

Consider edge (c, d)

c and d do not belong to the same tree. So, this edge can be added to A. Now, A has following elements:

{a}, {b}, {c, d, f}, {e}, {g}, {h}

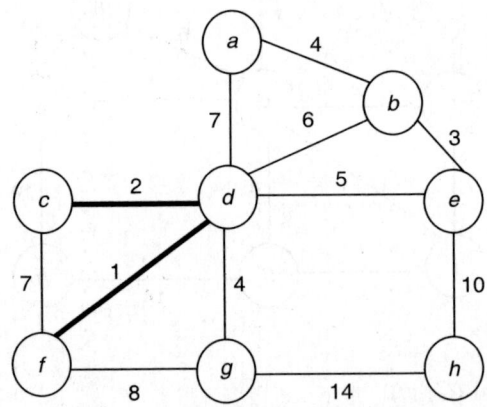

Consider edge (b, e)

b and e do not belong to the same tree so, add edge (b, e) to A.

Now, A has following elements:

{a}, {b, e}, {c, d, f}, {g}, {h}

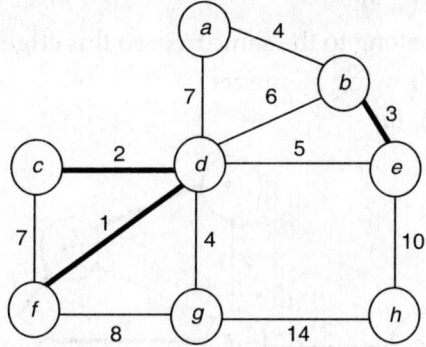

Consider edge $(a, b)$
$a$ and $b$ do not belong to the same tree, so, add this edge to A
A now contains following elements
$\{a, b, e\}, \{c, d, f\}, \{g\}, \{h\}$

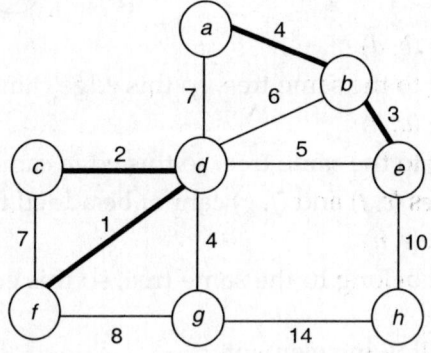

Consider edge $(d, g)$
$d$ and $g$ do not belong to the same tree, so add this edge to A.
Now, A has following elements:
$\{a, b, e\}, \{c, d, f, g\}, \{h\}$

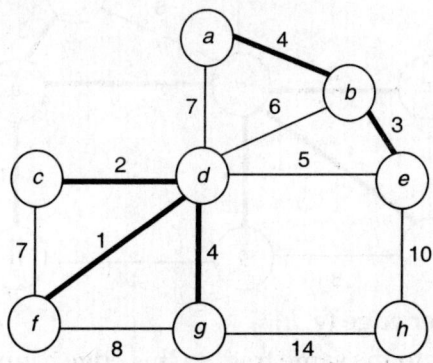

Consider edge (d, e)

d and e do not belong to the same tree, so this edge can be added to A

Now, A has following elements

{a, b, e, c, d, f, g}, {h}

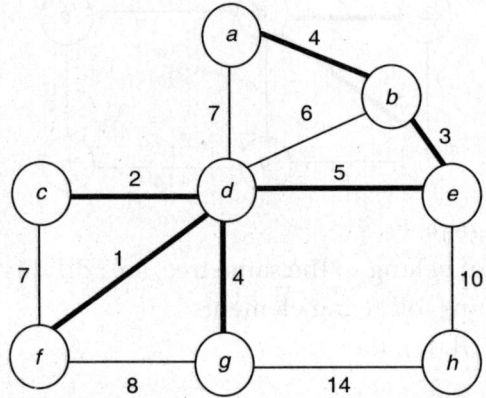

Consider edge (b, d)

b and d belong to the same tree, so this edge cannot be added to A

Consider edge (a, d)

a and d belong to the same tree, so this edge cannot be added to A

Similarly, edges (c, f) and (f, g) cannot be added to A

Consider edge (e, h)

e and h do not belong to the same tree, so this edge can be added to A

Now, A has following elements

{a, b, e, c, d, f, g, h }

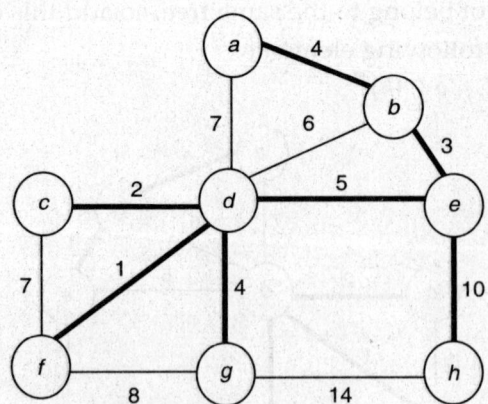

Finally consider edge (g, h)

g and h belong to the same tree, so this edge cannot be added to A.

### 3.2.4.2 Prim's algorithm

- Prim's algorithm is a minimum spanning tree algorithm.
- The Prim's algorithm finds a safe edge and adds it to the growing tree.
- Safe edge is an edge of least weight that connects the tree to an isolated vertex on which no edge of the tree is incident.

### Algorithm

MST-Prim (G, w, r)

| | |
|---|---|
| Line 1 | for each $u \in V[G]$ |
| Line 2 | key $[u] \leftarrow \infty$ |
| Line 3 | $\pi[u] \leftarrow$ nil |
| Line 4 | key $[r] \leftarrow 0$ |
| Line 5 | $Q \leftarrow V[G]$ |
| Line 6 | while $Q \neq \phi$ |
| Line 7 | $u \leftarrow$ Extract-Min (Q) |
| Line 8 | for each $v \in$ Adj $[u]$ |
| Line 9 | if $v \in Q$ and $w(u, v) <$ key $[v]$ |
| Line 10 | $\pi[v] \leftarrow u$ |
| Line 11 | key $[v] \leftarrow w(u, v)$ |

### Explanation

- The inputs to the above algorithm are a graph $G$ having weighted edges, $w$ denotes the weight of an edge and a source vertex $r$.
- Line 1 indicates the beginning of for loop that ends with Line 3. This loop is applicable for each vertex $u$ that belongs to the set of vertices of graph $G$.
- Line 2 indicates that the value of attribute key $[u]$ is set to $\infty$. In general the attribute key $[u]$ stores the minimum weight of any edge containing $u$ to a vertex in the tree. If key $[u]$ is set to $\infty$, it means no such edge exists.
- Line 3 indicates that the value of attribute $\pi[u]$ is set to nil. The attribute $\pi[u]$ stores the parent of vertex $u$.
- Line 4 indicates that the value of attribute key $[r]$ is set to 0. It shows that $r$ is the first vertex to be processed.
- Line 5 initializes the priority queue $Q$ by filling it with all the vertices of graph $G$.
- Line 6 indicates the beginning of while loop that ends with Line 11. This loop is applicable till the priority queue $Q$ becomes empty.

- Line 7 indicates that $u$ is the vertex extracted by the Extract-Min ($Q$) operation. The Extract-Min ($Q$) operation removes and returns the element of $Q$ with the smallest key.
- Line 8 indicates the beginning of for loop that ends with Line 11. This loop is applicable for each vertex $v$ that belongs to the adjacency list of vertex $u$.
- Line 9 checks the *if* condition. If this condition is true then the execution of Line 10 and Line 11 takes place. The *if* condition is true when vertex $v$ is an element of priority queue $Q$ and the weight of edge $(u, v)$ is less than the key of vertex $v$.
- Line 10 indicates that vertex $u$ becomes the parent of vertex $v$.
- Line 11 indicates that the weight of edge $(u, v)$ becomes the new value of attribute key $[v]$.

## Analysis

- The running time of Prim's algorithm depends on how we implement the min-priority queue $Q$.
- If we implement $Q$ as a binary min-heap, we can use the Build-Min-Heap procedure to perform Lines 1-5 in $O(|V|)$ time.
- The body of the while loop executes $|V|$ times, and since each Extract-Min operation takes $O(\lg |V|)$ time, the total time for all calls to Extract-Min is $O(|V| \lg |V|)$ time.
- The for loop in Lines 8-11 executes $O(|E|)$ times, since the sum of the lengths of all adjacency lists is $2|E|$.
- The assignment in Line 11 involves an implicit Decrease-key operation on the min-heap, which a binary min-heap supports in $O(\lg |V|)$ time.
- Thus, the total time for Prim's algorithm is $O(|V| \lg |V| + |E| \lg |V|) = O(|E| \lg |V|)$.

## Example

Show the result of Prim's algorithm on the following graph:

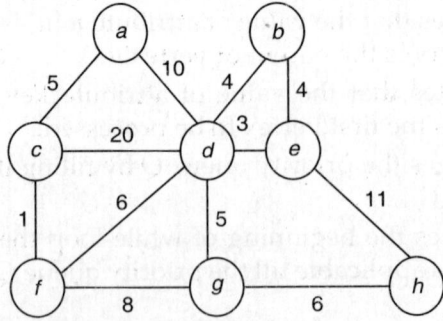

## Solution

MST-Prim $(G, w, a)$

After the execution of Line 5

| $u$ | $a$ | $b$ | $c$ | $d$ | $e$ | $f$ | $g$ | $h$ |
|---|---|---|---|---|---|---|---|---|
| $\pi[u]$ | nil | nil | nil | nil | nil | nil | nil | nil |
| key $[u]$ | 0 | $\infty$ | $\infty$ | $\infty$ | $\infty$ | $\infty$ | $\infty$ | $\infty$ |

$$Q \neq \phi$$

$u = a$ [Extract-Min $(Q)$ operation extracts the element $a$ having minimum key value zero]

Vertices adjacent to $a$ are $c$ and $d$.

for                     $v = c$

$c \in Q$ and $w(a, c) <$ key $[c]$

$c \in Q$ and $5 < \infty$

| $u$ | $b$ | $c$ | $d$ | $e$ | $f$ | $g$ | $h$ |
|---|---|---|---|---|---|---|---|
| $\pi[u]$ | nil | a | nil | nil | nil | nil | nil |
| key $[u]$ | $\infty$ | 5 | $\infty$ | $\infty$ | $\infty$ | $\infty$ | $\infty$ |

for                     $v = d$

$d \in Q$ and $w(a, d) <$ key $[d]$

$d \in Q$ and $10 < \infty$

| $u$ | $b$ | $c$ | $d$ | $e$ | $f$ | $g$ | $h$ |
|---|---|---|---|---|---|---|---|
| $\pi[u]$ | nil | a | a | nil | nil | nil | nil |
| key $[u]$ | $\infty$ | 5 | 10 | $\infty$ | $\infty$ | $\infty$ | $\infty$ |

$$Q \neq \phi$$

$u = c$ [Extract-Min $(Q)$ operation extracts the element $c$ having minimum key value 5]

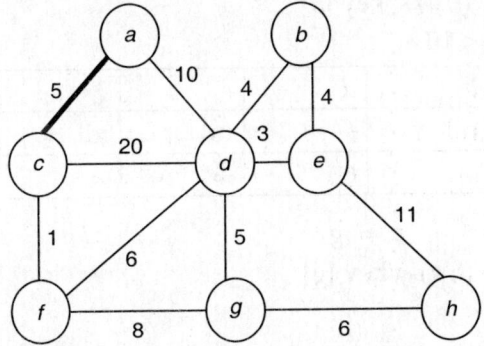

Vertices adjacent to $c$ are $a, d$ and $f$.

for                     $v = a$

$a \notin Q$

for                     $v = d$

$d \in Q$ but $w\,(c, d) \not< $ key $[d]$

$d \in Q$ but $20 \not< 10$

for                     $v = f$

$f \in Q$ and $w\,(c, f) < $ key $[f]$

$f \in Q$ and $1 < \infty$

| $u$ | $b$ | $d$ | $e$ | $f$ | $g$ | $h$ |
|---|---|---|---|---|---|---|
| $\pi\,[u]$ | nil | $a$ | nil | $c$ | nil | nil |
| key $[u]$ | $\infty$ | 10 | $\infty$ | 1 | $\infty$ | $\infty$ |

$$Q \neq \phi$$
$$u = f$$

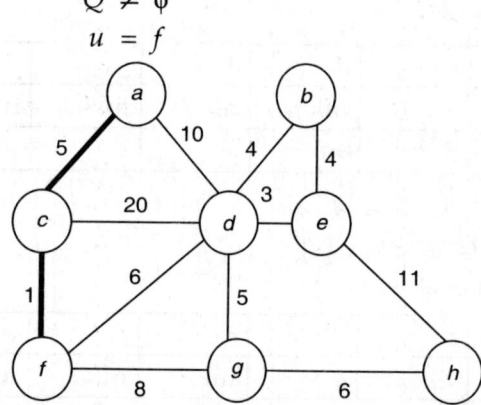

Vertices adjacent to $f$ are $c, d$ and $g$.

for                     $v = c$

$c \notin Q$

for                     $v = d$

$d \in Q$ and $w\,(f, d) < $ key $[d]$

$d \in Q$ and $6 < 10$

| $u$ | $b$ | $d$ | $e$ | $g$ | $h$ |
|---|---|---|---|---|---|
| $\pi\,[u]$ | nil | $f$ | nil | nil | nil |
| key $[u]$ | $\infty$ | 6 | $\infty$ | $\infty$ | $\infty$ |

for                     $v = g$

$g \in Q$ and $w\,(f, g) < $ key $[g]$

$g \in Q$ and $8 < \infty$

| u | b | d | e | g | h |
|---|---|---|---|---|---|
| $\pi[u]$ | nil | $f$ | nil | $f$ | nil |
| key $[u]$ | $\infty$ | 6 | $\infty$ | 8 | $\infty$ |

$$Q \neq \phi$$
$$u = d$$

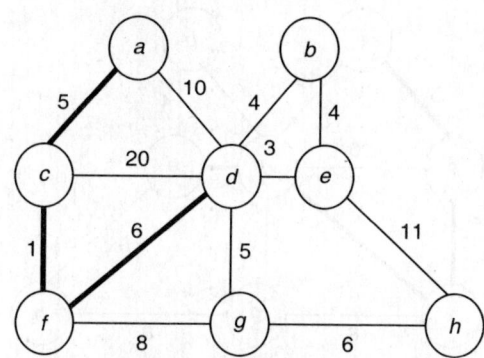

Vertices adjacent to $d$ are $a, c, f, b, g$ and $e$.

for $\qquad\qquad v = a$

$a \notin Q$

for $\qquad\qquad v = c$

$c \notin Q$

for $\qquad\qquad v = f$

$f \notin Q$

for $\qquad\qquad v = b$

$b \in Q$ and $w(d, b) < $ key $[b]$

$b \in Q$ and $4 < \infty$

| u | b | e | g | h |
|---|---|---|---|---|
| $\pi[u]$ | $d$ | nil | $f$ | nil |
| key $[u]$ | 4 | $\infty$ | 8 | $\infty$ |

for $\qquad\qquad v = g$

$g \in Q$ and $w(d, g) < $ key $[g]$

$g \in Q$ and $5 < 8$

| u | b | e | g | h |
|---|---|---|---|---|
| $\pi[u]$ | $d$ | nil | $d$ | nil |
| key $[u]$ | 4 | $\infty$ | 5 | $\infty$ |

$e \in Q$ and $w(d, e) < $ key $[e]$

$e \in Q$ and $3 < \infty$

| $u$ | $b$ | $e$ | $g$ | $h$ |
|---|---|---|---|---|
| $\pi[u]$ | $d$ | $d$ | $d$ | nil |
| key $[u]$ | 4 | 3 | 5 | $\infty$ |

$$Q \neq \phi$$
$$u = e$$

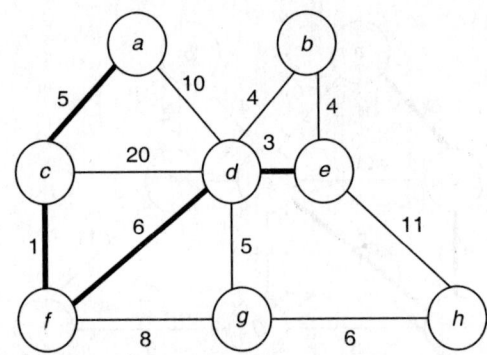

Vertices adjacent to $e$ are $b$, $d$ and $h$.
$$v = b$$
$b \in Q$ but $w(e, b) \not< $ key $[b]$
$b \in Q$ but $4 \not< 4$
$$v = d$$
$d \notin Q$
$$v = h$$
$h \in Q$ and $w(e, h) <$ key $[h]$
$h \in Q$ and $11 < \infty$

| $u$ | $b$ | $g$ | $h$ |
|---|---|---|---|
| $\pi[u]$ | $d$ | $d$ | $e$ |
| key $[u]$ | 4 | 5 | 11 |

$$Q \neq \phi$$
$$u = b$$

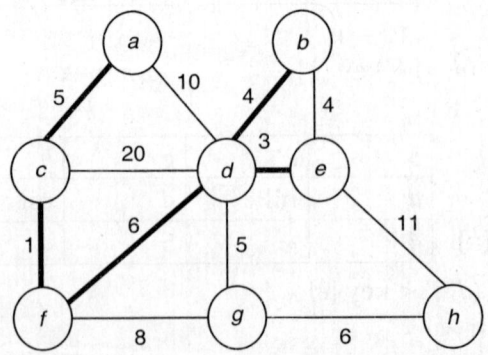

Vertices adjacent to *b* are *d* and *e*.

for                    $v = d$

$d \notin Q$

for                    $v = e$

$e \notin Q$

| $u$ | $g$ | $h$ |
|---|---|---|
| $\pi [u]$ | $d$ | $e$ |
| key [$u$] | 5 | 11 |

$Q \neq \phi$

$u = g$

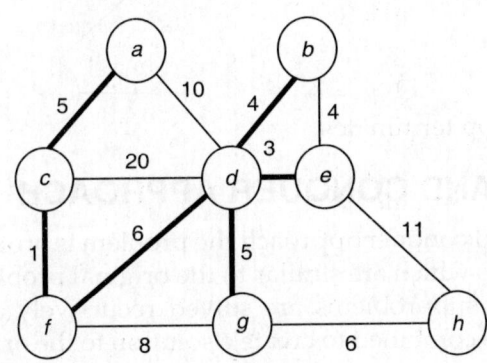

Vertices adjacent to *g* are *f*, *d* and *h*.

for                    $v = f$

$f \notin Q$

for                    $v = d$

$d \notin Q$

for                    $v = h$

$h \in Q$ and $w (g, h) < $ key [$h$]

$h \in Q$ and $6 < 11$

| $u$ | $h$ |
|---|---|
| $\pi [u]$ | g |
| key [$u$] | 6 |

$Q \neq \phi,$

$u = h$

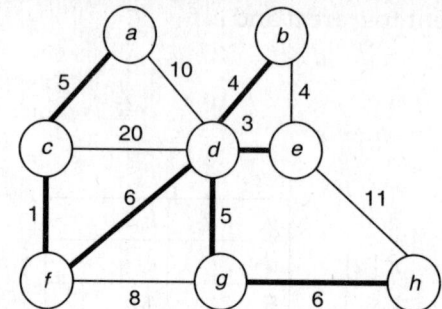

Vertices adjacent to $h$ are $e$ and $g$

for                        $v = e$

$e \notin Q$

for                        $v = g$

$g \notin Q$

Now                    $Q = = \phi$

So, while loop terminates.

## 3.3 DIVIDE AND CONQUER APPROACH

- In divide and conquer approach the problem is broken into several subproblems which are similar to the original problem but smaller in size. The subproblems are solved recursively and then these solutions are combined to create a solution to the original problem.
- The divide and conquer approach involves three steps at each level of the recursion:
  - (i) Divide the problem into a number of subproblems that are smaller instances of the same problem.
  - (ii) Conquer the subproblems by solving them recursively.
  - (iii) Combine the solutions to the subproblems into the solution for the original problem.
- Divide and conquer approach is a basis for many kinds of problems like sorting (e.g. merge sort, quick sort), multiplying large numbers, syntactic analysis (e.g. top down parsers) etc.

### 3.3.1 Quick Sort

**Algorithm**

```
        Quick sort (A, p, r)
Line 1        if p < r
Line 2            q ← Partition (A, p, r)
Line 3            Quick sort (A, p, q – 1)
Line 4            Quick sort (A, q + 1, r)
```

## Explanation

- The inputs to the Quick sort algorithm are an array A, a pointer '$p$' to the first element of array A, and a pointer '$r$' to the last element of array A.
- Line 1 checks the *if* condition. If this condition is true then the execution of Line 2 to Line 4 takes place else the procedure terminates.
- Line 2 indicates that $q$ is the pointer returned by the procedure Partition (A, $p$, $r$).

  This procedure Partition (A, $p$, $r$) partitions the array A [$p...r$] into two subarrays A [$p...q-1$] and A [$q+1...r$] such that each element of A [$p...q-1$] is less than or equal to A[$q$] which is, in turn less than or equal to each element of A [$q+1...r$].
- Line 3 calls the procedure Quick sort (A, $p$, $q-1$) to sort the subarray A [$p...q-1$].
- Line 4 calls the procedure Quick sort (A, $q+1$, $r$) to sort the subarray A [$q+1...r$].

## Algorithm

        Partition (A, p, r)
Line 1       $x \leftarrow A [r]$
Line 2       $i \leftarrow p - 1$
Line 3       for $j \leftarrow p$ to $r - 1$
Line 4            if A[j] $\leq$ x
Line 5                 $i \leftarrow i + 1$
Line 6                     exchange A [i] with A[j]
Line 7            exchange A [i + 1] with A [r]
Line 8       return i + 1

## Explanation

- Line 1 indicates that $x$ is an array element at index $r$ and $x$ is a pivot element. The partition procedure partitions the sub-array A [$p...r$] into four regions:
  (i) The values in A [$p...i$] are all less than or equal to $x$.
  (ii) The values in A [$i + 1...j - 1$] are all greater than $x$.
  (iii) The sub-array A [$j...r - 1$] can take any values.
  (iv) The value A[$r$] is equal to $x$.

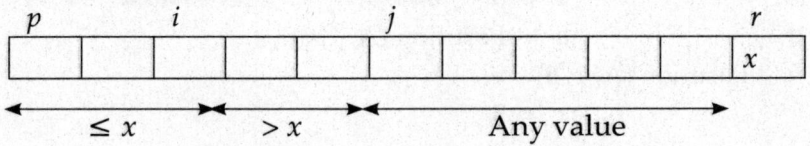

- Line 2 indicates that index $i$ is one less than index $p$.
- Line 3 indicates the beginning of for loop that ends with Line 6. This for loop is applicable for $j$ equals to $p$ to $r - 1$.
- Line 4 checks the *if* condition. If this condition is true then the execution of Line 5 and Line 6 takes place else the control goes back to Line 3.
- The *if* condition is true when the array element at location $j$ is less than or equal to the pivot element $x$.
- Line 5 increments the value of $i$ by 1.
- Line 6 exchanges the element A $[i]$ with A $[j]$.
- Line 7 exchanges the element A $[i + 1]$ with A$[r]$.
- Line 8 returns the index $i + 1$.

## Analysis

- The worst case running time of Quick sort occurs when the partitioning procedure produces most unbalanced partitioning. This happens when the procedure produces one subproblem with $n - 1$ elements and the other with 0 elements.
- Assume that this unbalanced partitioning arises in each recursive call. The partitioning costs is $\Theta(n)$ time. Since the recursive call on an array of size 0 just returns, $T(0) = \Theta(1)$.
- Following recurrence relation describes the worst case running time of Quick sort.

$$T(n) = T(n-1) + T(0) = \Theta(n)$$
$$= T(n-1) + \Theta(n)$$

The solution to the above recurrence is

$$T(n) = \Theta(n^2).$$

## Example

Illustrate the operation of Quicksort (A, $p$, $r$) algorithm on the following array.

|   | 1 | 2 | 3 | 4 | 5 | 6 |
|---|---|---|---|---|---|---|
| A = | 2 | 5 | 4 | 1 | 2 | 3 |

## Solution

Quick sort (A, 1, 6)

$$1 < 6$$
$$q = \text{Partition (A, 1, 6)}$$

Partition (A, 1, 6)

$$x = 3$$

$$i = 0$$
$$j = 1 \text{ to } 5$$
for                     $j = 1$
$$2 \leq 3$$
$$i = 1$$

| 1 | 2 | 3 | 4 | 5 | 6 |
|---|---|---|---|---|---|
| 2 | 5 | 4 | 1 | 2 | 3 |

for                     $j = 2$
$$5 \nleq 3$$
for                     $j = 3$
$$4 \nleq 3$$
for                     $j = 4$
$$1 \leq 3$$
$$i = 2$$

| 1 | 2 | 3 | 4 | 5 | 6 |
|---|---|---|---|---|---|
| 2 | 1 | 4 | 5 | 2 | 3 |

for                     $j = 5$
$$2 \leq 3$$
$$i = 3$$

| 1 | 2 | 3 | 4 | 5 | 6 |
|---|---|---|---|---|---|
| 2 | 1 | 2 | 5 | 4 | 3 |

| 1 | 2 | 3 | 4 | 5 | 6 |
|---|---|---|---|---|---|
| 2 | 1 | 2 | 3 | 4 | 5 |

return 4

So,                 $q = 4$

Quicksort (A, 1, 3)

$$1 < 3$$
$$q = \text{Partition (A, 1, 3)}$$

Partition (A, 1, 3)

$$x = 2$$
$$i = 0$$
$$j = 1 \text{ to } 2$$
for                     $j = 1$
$$2 \leq 2$$
$$i = 1$$

| 1 | 2 | 3 | 4 | 5 | 6 |
|---|---|---|---|---|---|
| 2 | 1 | 2 | 3 | 4 | 5 |

for                $j = 2$
                   $1 \leq 2$
                   $i = 2$

| 1 | 2 | 3 | 4 | 5 | 6 |
|---|---|---|---|---|---|
| 2 | 1 | 2 | 3 | 4 | 5 |

| 1 | 2 | 3 | 4 | 5 | 6 |
|---|---|---|---|---|---|
| 2 | 1 | 2 | 3. | 4 | 5 |

return 3

So,                $q = 3$

Quick sort (A, 1, 2)

                   $1 < 2$
                   $q = $ Partition (A, 1, 2)

Partition (A, 1,2)

                   $x = 1$
                   $i = 0$
                   $j = 1$

for                $j = 1$
                   $2 \nleq 1$

| 1 | 2 | 3 | 4 | 5 | 6 |
|---|---|---|---|---|---|
| 1 | 2 | 2 | 3 | 4 | 5 |

return 1

   So,             $q = 1$

Quick sort (A, 1, 0)

                   $1 \nless 0$

Quick sort (A, 2, 2)

                   $2 \nless 2$

Quick sort (A, 4, 3)

                   $4 \nless 3$

Quick sort (A, 5, 6)

                   $5 < 6$
                   $q = $ Partition (A, 5, 6)

Partition (A, 5, 6)

                   $x = 5$

$$i = 4$$
$$j = 5$$
$$4 \leq 5$$
$$i = 5$$

| 1 | 2 | 3 | 4 | 5 | 6 |
|---|---|---|---|---|---|
| 1 | 2 | 2 | 3 | 4 | 5 |

| 1 | 2 | 3 | 4 | 5 | 6 |
|---|---|---|---|---|---|
| 1 | 2 | 2 | 3 | 4 | 5 |

return 6

So,                    $q = 6$

Quick sort (A, 5, 5)

$$5 \not< 5$$

Now, procedure terminates.

# 3.4 DECREASE AND CONQUER APPROACH (INCREMENTAL APPROACH)

- Decrease and conquer approach involves the following three steps:
  - (i) Reduce problem instance to smaller instance of the same problem.
  - (ii) Solve smaller instance.
  - (iii) Extend solution of smaller instance to obtain solution to original instance.
- Decrease and conquer approach can be used in insertion sort, graph search algorithms (like breadth first search and depth first search) etc.
- In insertion sort, breadth first search and depth first search, the problem is decreased by a constant (usually by 1).

## 3.4.1 Insertion Sort

- It is an efficient algorithm for sorting a small number of elements.
- Input: Sequence of $n$ numbers.
- Output: Sorted sequence such that the $n$ numbers are arranged in increasing order.
- The time taken by the insertion sort procedure depends on the input. Greater the size of sequence more will be the sorting time.
- Insertion sort can take different amount of time to sort two input sequences of the same size depending on how nearly sorted they already are.

## Algorithm

Insertion sort (A)

Line 1  for j ← 2 to length [A]
Line 2          key ← A [j]
Line 3          i ← j – 1
Line 4     while i > 0 and A[i] > key
Line 5          A [i + 1] ← A[i]
Line 6          i ← i – 1
Line 7     A [i + 1] ← key

## Explanation

- The body of the for loop begins with Line 1 and ends with Line 7.
- The body of the while loop begins with Line 4 and ends with Line 6.
- length [A] indicates the number of elements in array A. length [A] = n
- key is a variable.
- i and j are the array indices.
- Line 1 indicates the beginning of for loop. The array index j is initialized as 2. The value of j increments by 1 each time for the next iteration of the for loop. j can take values from 2 to n. The for loop terminates when j > n.
- Line 2 indicates that the variable 'key' holds the value stored in the array at location j.
- Line 3 indicates that the index i has a value 1 less than index j.
- Line 4 indicates the beginning of while loop. This loop repeats itself until both the conditions are satisfied. The loop terminates itself as soon as one of the conditions fails.

  1st condition: The value of i should always be greater than 0.

  2nd condition: The value stored in the array at location i should be greater than value stored in the variable 'key'.
- Line 5 indicates that in the array the value at location i + 1 is replaced by the value at location i.
- Line 6 indicates that the index i is decremented by 1.
- Line 7 indicates that the location i + 1 gets the value stored in the variable key.

## Analysis

- The worst case running time of insertion sort is $\Theta$ $(n^2)$.

## Example

Apply insertion sort on the given sequence of numbers 5, 3, 4, 1.

## Solution

| 1 | 2 | 3 | 4 |
|---|---|---|---|
| 5 | 3 | 4 | 1 |

Indices of the array are shown above the boxes.

Here 1, 2, 3, 4 are the indices.

Values stored in the array at positions pointed by indices are shown within the boxes.

Here 5, 3, 4, 1 are the values stored at locations pointed by indices 1, 2, 3, 4 respectively.

$$j = 2$$
$$key = 3$$
$$i = 1$$

$1 > 0$ and $5 > 3$

| 1 | 2 | 3 | 4 |
|---|---|---|---|
| 5 | 5 | 4 | 1 |

$$i = 0$$
$$0 \not> 0$$

| 1 | 2 | 3 | 4 |
|---|---|---|---|
| 3 | 5 | 4 | 1 |

$$j = 3$$
$$key = 4$$
$$i = 2$$

$2 > 0$ and $5 > 4$

| 1 | 2 | 3 | 4 |
|---|---|---|---|
| 3 | 5 | 5 | 1 |

$$i = 1$$

$1 > 0$ and $3 \not> 4$

| 1 | 2 | 3 | 4 |
|---|---|---|---|
| 3 | 4 | 5 | 1 |

$$j = 4$$
$$key = 1$$
$$i = 3$$

$3 > 0$ and $5 > 1$

| 1 | 2 | 3 | 4 |
|---|---|---|---|
| 3 | 4 | 5 | 5 |

$$i = 2$$

$2 > 0$ and $4 > 1$

| 1 | 2 | 3 | 4 |
|---|---|---|---|
| 3 | 4 | 4 | 5 |

$$i = 1$$

$1 > 0$ and $3 > 1$

| 1 | 2 | 3 | 4 |
|---|---|---|---|
| 3 | 3 | 4 | 5 |

$$i = 0$$
$$0 \ngtr 0$$

| 1 | 2 | 3 | 4 |
|---|---|---|---|
| 1 | 3 | 4 | 5 |

At $\quad\quad\quad\quad j = 5$, the for loop terminates.

### Example

Apply insertion sort on the given sequence of numbers 8, 5, 9, 1, 4, 3, 2.

### Solution

| 1 | 2 | 3 | 4 | 5 | 6 | 7 |
|---|---|---|---|---|---|---|
| 8 | 5 | 9 | 1 | 4 | 3 | 2 |

$$j = 2$$
$$key = 5$$
$$i = 1$$

$i > 0$ and $8 > 5$

| 1 | 2 | 3 | 4 | 5 | 6 | 7 |
|---|---|---|---|---|---|---|
| 8 | 8 | 9 | 1 | 4 | 3 | 2 |

$$i = 0$$
$$0 \ngtr 0$$

| 1 | 2 | 3 | 4 | 5 | 6 | 7 |
|---|---|---|---|---|---|---|
| 5 | 8 | 9 | 1 | 4 | 3 | 2 |

$$j = 3$$
$$\text{key} = 9$$
$$i = 2$$

2 > 0 and 8 ≯ 9

| 1 | 2 | 3 | 4 | 5 | 6 | 7 |
|---|---|---|---|---|---|---|
| 5 | 8 | 9 | 1 | 4 | 3 | 2 |

$$j = 4$$
$$\text{key} = 1$$
$$i = 3$$

3 > 0 and 9 > 1

| 1 | 2 | 3 | 4 | 5 | 6 | 7 |
|---|---|---|---|---|---|---|
| 5 | 8 | 9 | 9 | 4 | 3 | 2 |

$$i = 2$$

2 > 0 and 8 > 1

| 1 | 2 | 3 | 4 | 5 | 6 | 7 |
|---|---|---|---|---|---|---|
| 5 | 8 | 8 | 9 | 4 | 3 | 2 |

$$i = 1$$

1 > 0 and 5 > 1

| 1 | 2 | 3 | 4 | 5 | 6 | 7 |
|---|---|---|---|---|---|---|
| 5 | 5 | 8 | 9 | 4 | 3 | 2 |

$$i = 0$$

0 ≯ 0

| 1 | 2 | 3 | 4 | 5 | 6 | 7 |
|---|---|---|---|---|---|---|
| 1 | 5 | 8 | 9 | 4 | 3 | 2 |

$$j = 5$$
$$\text{key} = 4$$
$$i = 4$$

4 > 0 and 9 > 4

| 1 | 2 | 3 | 4 | 5 | 6 | 7 |
|---|---|---|---|---|---|---|
| 1 | 5 | 8 | 9 | 9 | 3 | 2 |

$$i = 3$$

3 > 0 and 8 > 4

| 1 | 2 | 3 | 4 | 5 | 6 | 7 |
|---|---|---|---|---|---|---|
| 1 | 5 | 8 | 8 | 9 | 3 | 2 |

$$i = 2$$

$2 > 0$ and $5 > 4$

| 1 | 2 | 3 | 4 | 5 | 6 | 7 |
|---|---|---|---|---|---|---|
| 1 | 5 | 5 | 8 | 9 | 3 | 2 |

$$i = 1$$

$1 > 0$ and $1 \not> 4$

| 1 | 2 | 3 | 4 | 5 | 6 | 7 |
|---|---|---|---|---|---|---|
| 1 | 4 | 5 | 8 | 9 | 3 | 2 |

$$j = 6$$
$$key = 3$$
$$i = 5$$

$5 > 0$ and $9 > 3$

| 1 | 2 | 3 | 4 | 5 | 6 | 7 |
|---|---|---|---|---|---|---|
| 1 | 4 | 5 | 8 | 9 | 9 | 2 |

$$i = 4$$

$4 > 0$ and $8 > 3$

| 1 | 2 | 3 | 4 | 5 | 6 | 7 |
|---|---|---|---|---|---|---|
| 1 | 4 | 5 | 8 | 8 | 9 | 2 |

$$i = 3$$

$3 > 0$ and $5 > 3$

| 1 | 2 | 3 | 4 | 5 | 6 | 7 |
|---|---|---|---|---|---|---|
| 1 | 4 | 5 | 5 | 8 | 9 | 2 |

$$i = 2$$

$2 > 0$ and $4 > 3$

| 1 | 2 | 3 | 4 | 5 | 6 | 7 |
|---|---|---|---|---|---|---|
| 1 | 4 | 4 | 5 | 8 | 9 | 2 |

$$i = 1$$

$1 > 0$ and $1 \not> 3$

| 1 | 2 | 3 | 4 | 5 | 6 | 7 |
|---|---|---|---|---|---|---|
| 1 | 3 | 4 | 5 | 8 | 9 | 2 |

$$j = 7$$
$$key = 2$$
$$i = 6$$

6 > 0 and 9 > 2

| 1 | 2 | 3 | 4 | 5 | 6 | 7 |
|---|---|---|---|---|---|---|
| 1 | 3 | 4 | 5 | 8 | 9 | 9 |

$$i = 5$$

5 > 0 and 8 > 2

| 1 | 2 | 3 | 4 | 5 | 6 | 7 |
|---|---|---|---|---|---|---|
| 1 | 3 | 4 | 5 | 8 | 8 | 9 |

$$i = 4$$

4 > 0 and 5 > 2

| 1 | 2 | 3 | 4 | 5 | 6 | 7 |
|---|---|---|---|---|---|---|
| 1 | 3 | 4 | 5 | 5 | 8 | 9 |

$$i = 3$$

3 > 0 and 4 > 2

| 1 | 2 | 3 | 4 | 5 | 6 | 7 |
|---|---|---|---|---|---|---|
| 1 | 3 | 4 | 4 | 5 | 8 | 9 |

$$i = 2$$

2 > 0 and 3 > 2

| 1 | 2 | 3 | 4 | 5 | 6 | 7 |
|---|---|---|---|---|---|---|
| 1 | 3 | 3 | 4 | 5 | 8 | 9 |

$$i = 1$$

1 > 0 and 1 ≯ 2

| 1 | 2 | 3 | 4 | 5 | 6 | 7 |
|---|---|---|---|---|---|---|
| 1 | 2 | 3 | 4 | 5 | 8 | 9 |

At $j = 8$, the for loop terminates.

## 3.4.2 Breadth First Search (BFS)

Consider a graph $G = (V, E)$

where,   $V$ represents vertices of graph $G$

$E$ represents edges of graph $G$

- Breadth first search algorithm searches every vertex $v$ which is reachable from the source vertex $s$.

- BFS is used to find the shortest path from source vertex $s$ to each reachable vertex $v$, $s$ and $v \in V$.
- For BFS this shortest path distance is the minimum number of edges in any path from source vertex $s$ to vertex $v$.
- BFS also produces a 'breadth first tree' that contains all the vertices reachable from the source vertex.
- Initially this tree contains only the source vertex.
- Whenever, the search discovers a white vertex $v$ from the adjacency list of an already discovered vertex $u$, the vertex $v$ and the edge $(u, v)$ are added to the tree.
- Breadth first tree may vary because it depends upon the order in which the neighbors of a given vertex are visited in Line 12 of the algorithm.

  But the distances $d$ computed by the algorithm will be same for all cases.
- White color is assigned to all the undiscovered nodes.
- Gray color is assigned to all the discovered but unvisited nodes.
- Black color is assigned to all the visited nodes.
- BFS is applicable for both directed and undirected graphs.
- BFS is so named because the algorithm discovers all vertices at a distance $k$ from source vertex $s$ before discovering any vertices at distance $k + 1$.

## Algorithm

    BFS (G, s)

| | |
|---|---|
| Line 1 | for each vertex u $\in$ V [G] – {s} |
| Line 2 | c [u] $\leftarrow$ white |
| Line 3 | d [u] $\leftarrow \infty$ |
| Line 4 | $\pi$ [u] $\leftarrow$ nil |
| Line 5 | c [s] $\leftarrow$ gray |
| Line 6 | d [s] $\leftarrow$ 0 |
| Line 7 | $\pi$ [s] $\leftarrow$ nil |
| Line 8 | Q $\leftarrow \phi$ |
| Line 9 | Enqueue (Q, s) |
| Line 10 | while Q $\neq \phi$ |
| Line 11 | u $\leftarrow$ Dequeue (Q) |
| Line 12 | for each vertex v $\in$ adj [u] |
| Line 13 | if c[v] = = white |
| Line 14 | c [v] $\leftarrow$ gray |

| Line 15 | d [v] ← d [u] + 1 |
| Line 16 | π [v] ← u |
| Line 17 | Enqueue (Q, v) |
| Line 18 | c [u] ← black |

## Explanation

- Line 1 indicates the beginning of for loop that ends with Line 4. G represents a graph.

  s represents the source vertex.

  V [G] represents the set of vertices of graph G.

  This for loop is applicable for each of the vertex that belongs to the set of vertices of graph G except the source vertex s.
- Line 2 indicates that white color is assigned to vertex u.

  Attribute c[u] stores the color of the vertex u.
- Line 3 indicates that the distance from source s to vertex u is infinity (no path from s to u).

  Attribute d[u] stores the distance from the source vertex s to vertex u computed by the algorithm.
- Line 4 indicates that the parent of vertex u is nil.

  Attribute π [u] stores the parent of vertex u.
- Line 5 indicates that the gray color is assigned to source vertex s.
- Line 6 indicates that the distance from source s to itself is zero.
- Line 7 indicates that the parent of source vertex s is nil.
- Line 8 indicates that as a data structure a first-in-first-out queue Q has been taken which is initially empty.

  This queue Q has been used to manage the gray vertices.
- Line 9 indicates that queue Q has been initialized by entering a source vertex s into it.

  Enqueue (Q) is the operation of filling the Q by inserting vertices into it.
- Line 10 indicates the beginning of while loop that ends with Line 18. This while loop terminates when the queue Q becomes empty.
- Line 11 indicates that u is that vertex which is present at the head of the queue Q and removes from there.

  Dequeue (Q) is the operation of emptying the Q by removing vertices from it.
- Line 12 indicates the beginning of for loop that ends with Line 17.

  This for loop is applicable for each vertex v that belongs to the adjacency list of u.

  adj [u] represents the adjacency list of u.

- Line 13 indicates the checking of *if* condition.

  This condition is used to check whether the color of vertex $v$ is white.

  If the condition is true then the execution of Line 14 to Line 17 takes place.
- Line 14 indicates that gray color is assigned to vertex $v$.
- Line 15 indicates that the distance from source $s$ to vertex $v$ is equal to one more than the distance from source $s$ to vertex $u$.
- Line 16 indicates that the parent of $v$ is $u$.
- Line 17 indicates the entering of vertex $v$ into the queue $Q$. This vertex enters from the tail of the queue $Q$.
- Line 18 indicates that the black color is assigned to vertex $u$.

**Analysis**

- Time taken for initialization is $O\,(|V|)$.
- After initialization, breadth first search never whitens a vertex and thus, the test in Line 13 ensures that each vertex is enqueued at most once and hence dequeued at most once. The operations of enqueuing and dequeuing take $O\,(1)$ time. So, the total time taken for queue operations is $O\,(|V|)$.
- Because the procedure scans the adjacency list of each vertex only when the vertex is dequeued, it scans each adjacency list at most once. Since, the sum of lengths of all the adjacency list is $\Theta\,(|E|)$, the total time spent in scanning adjacency lists is $O\,(|E|)$.
- Thus, the total running time of the BFS procedure is $O\,(|V| + |E|)$.

**Example**

Apply BFS operation on the given graph, consider $v_1$ to be the source vertex.

## Solution

The following figure is obtained after the execution of first for loop.

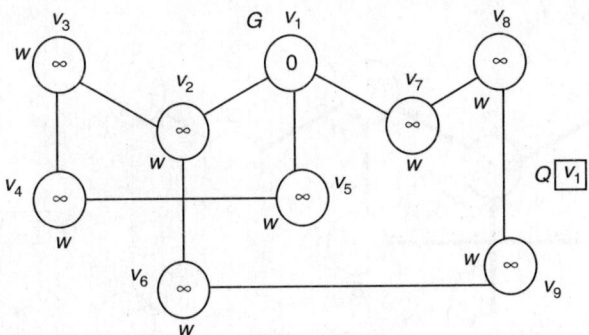

- The value of attribute $d[u]$ is appeared within the vertex.
- $W$, $G$, $B$ represents the white, gray, and black colors respectively.
- A data structure queue $Q$ is also maintained.
- $u = v_1$ {Execution of while loop}
  Vertices adjacent to $v_1$ are $v_2$, $v_5$ and $v_7$.
  For all these vertices, if condition is true.

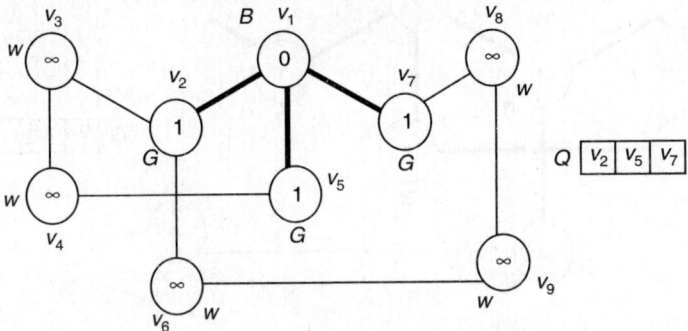

$u = v_2$

Vertices adjacent to $v_2$ are $v_1$, $v_3$ and $v_6$, out of which *if* condition is true for vertices $v_3$ and $v_6$.

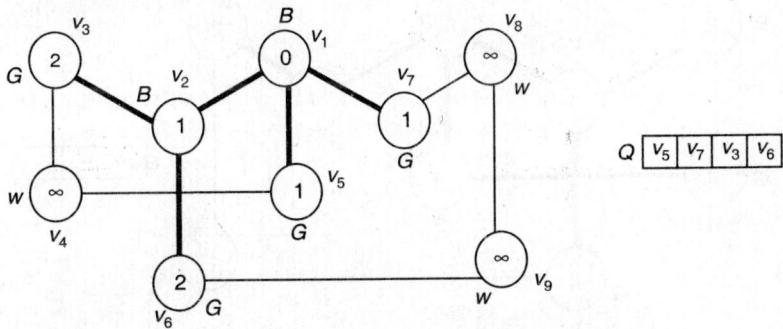

$u = v_5$

Vertices adjacent to $v_5$ are $v_1$ and $v_4$, out of which if condition is true for vertex $v_4$.

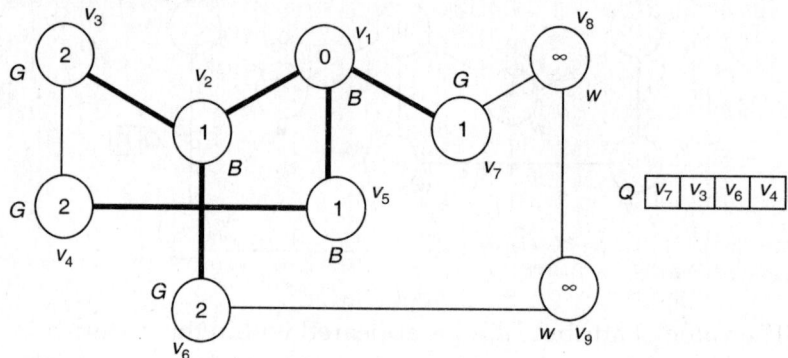

$u = v_7$

Vertices adjacent to $v_7$ are $v_1$ and $v_8$, out of which if condition is true for vertex $v_8$.

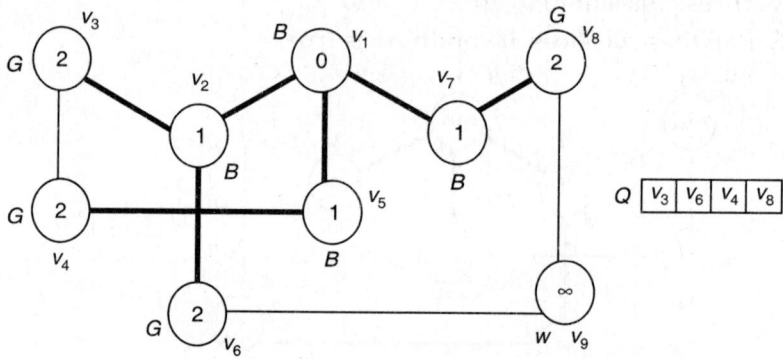

$u = v_3$

Vertices adjacent to $v_3$ are $v_2$ and $v_4$, if condition is not true for any of them.

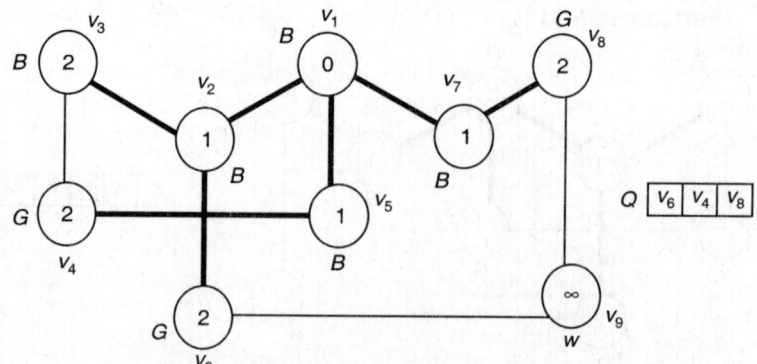

$u = v_6$

Vertices adjacent to $v_6$ are $v_2$ and $v_9$, if condition is true for $v_9$.

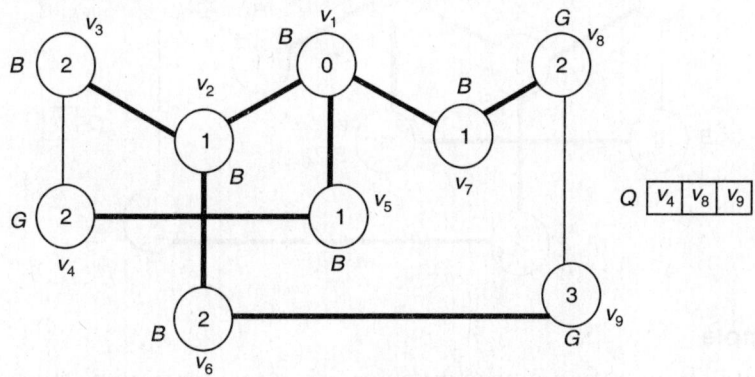

$u = v_4$

Vertices adjacent to $v_4$ are $v_3$ and $v_5$, if condition is not true for any of them.

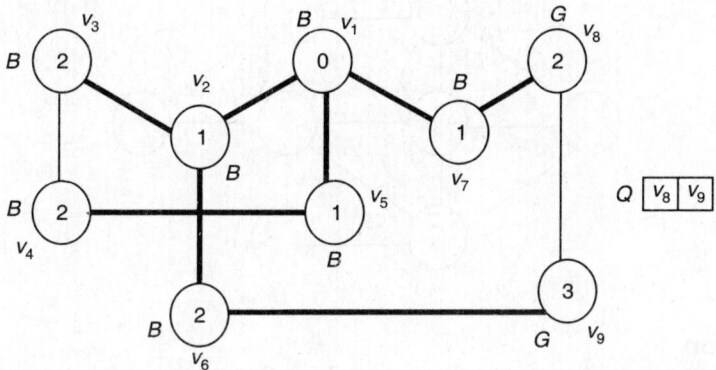

$u = v_8$

Vertices adjacent to $v_8$ are $v_7$ and $v_9$, if condition is not true for both of them.

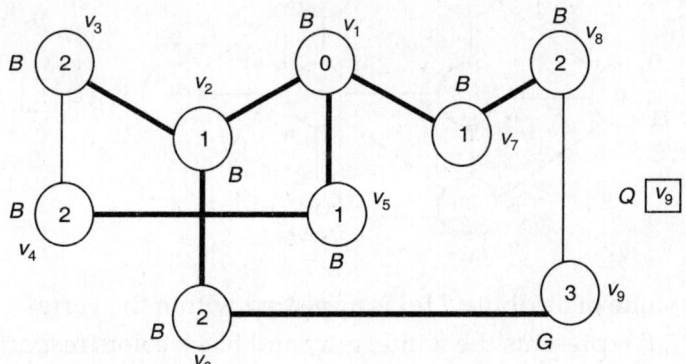

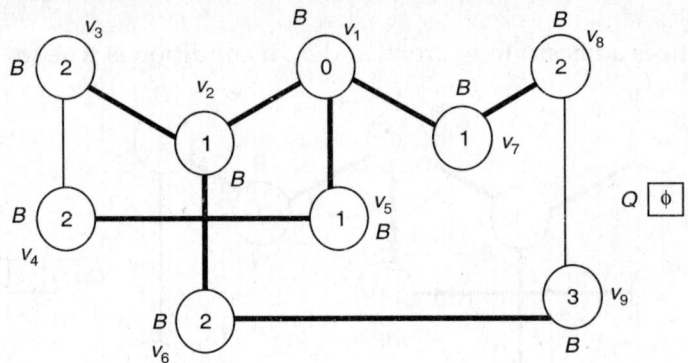

## Example

Apply BFS operation on the given graph, consider $v_1$ to be the source vertex.

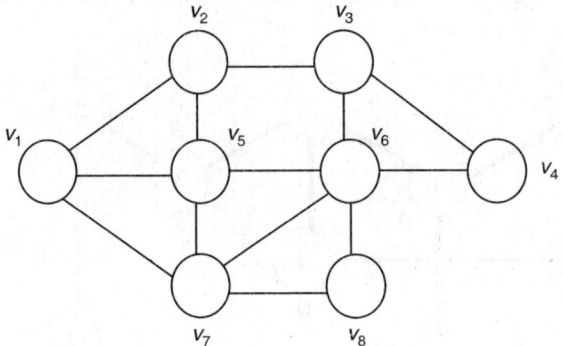

## Solution

The following figure is obtained after the execution of first for loop.

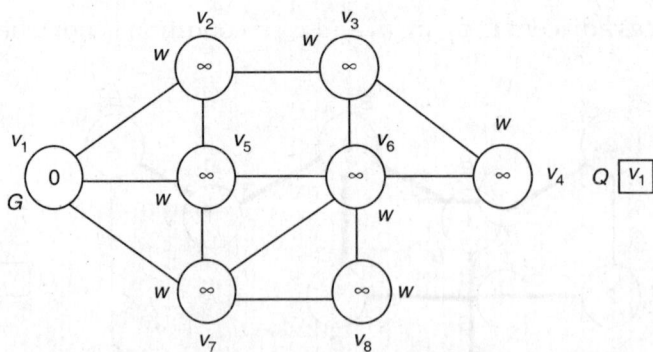

- The value of attribute $d[u]$ is appeared within the vertex.
- $W$, $G$, $B$ represents the white, gray and black colors respectively.

- A data structure queue $Q$ is also maintained.

  Execution of while loop,

  $u = v_1$

  Vertices adjacent to $v_1$ are $v_2$, $v_5$ and $v_7$, if condition is true for all of them.

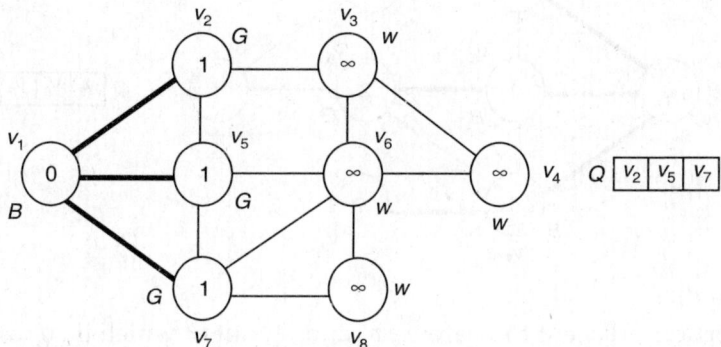

$u = v_2$

Vertices adjacent to $v_2$ are $v_1$, $v_3$ and $v_5$, out of which if condition is true for vertex $v_3$.

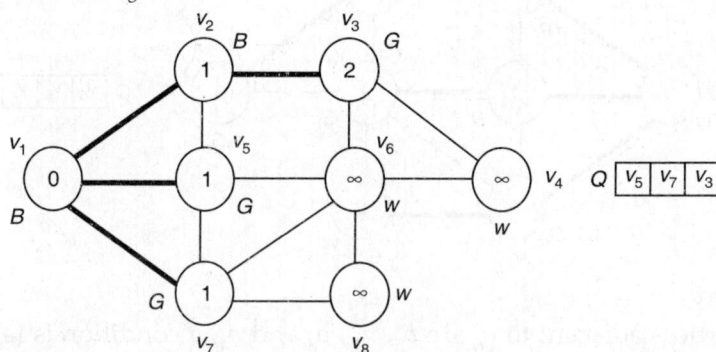

$u = v_5$

Vertices adjacent to $v_5$ are $v_1$, $v_2$, $v_6$ and $v_7$, out of which if condition is true for vertex $v_6$.

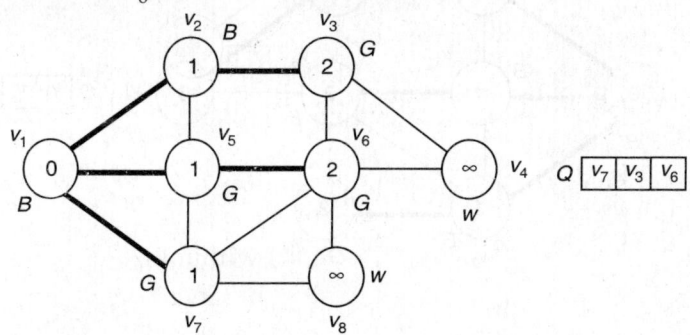

$u = v_7$

Vertices adjacent to $v_7$ are $v_1, v_5, v_6$ and $v_8$, out of which if condition is true for vertex $v_8$.

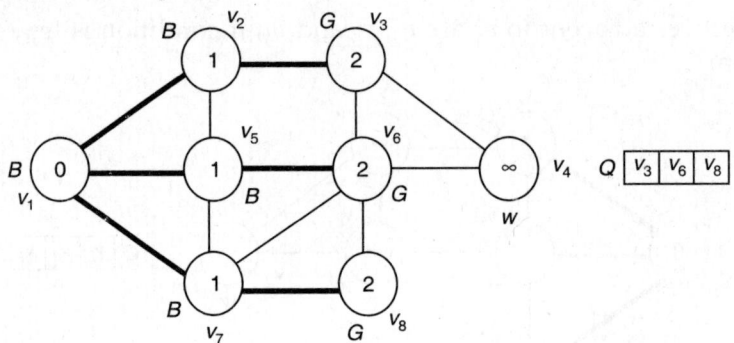

$u = v_3$

Vertices adjacent to $v_3$ are $v_2, v_6$ and $v_4$, out of which if condition is true for vertex $v_4$.

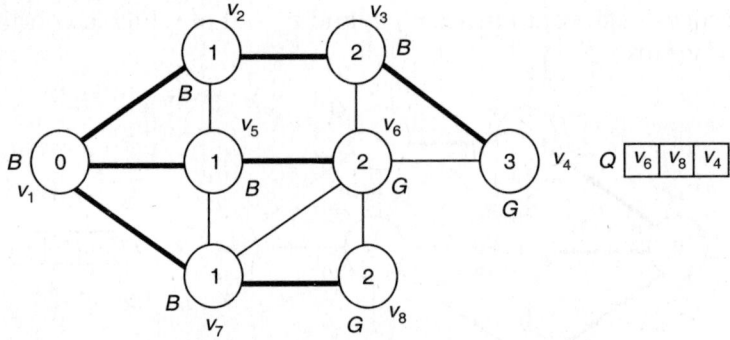

$u = v_6$

Vertices adjacent to $v_6$ are $v_5, v_3, v_8$ and $v_4$, if condition is false for all of them.

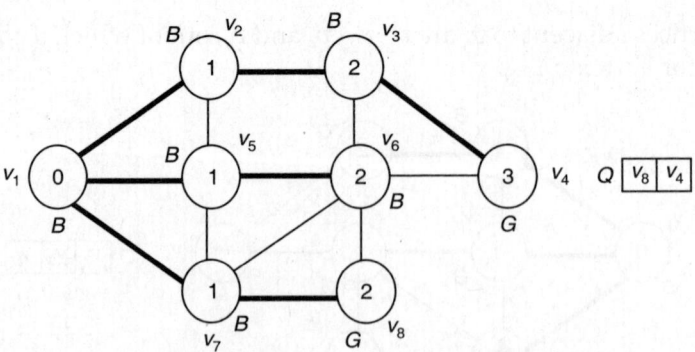

$u = v_8$

Vertices adjacent to $v_8$ are $v_6$ and $v_7$, if condition is false for both of them.

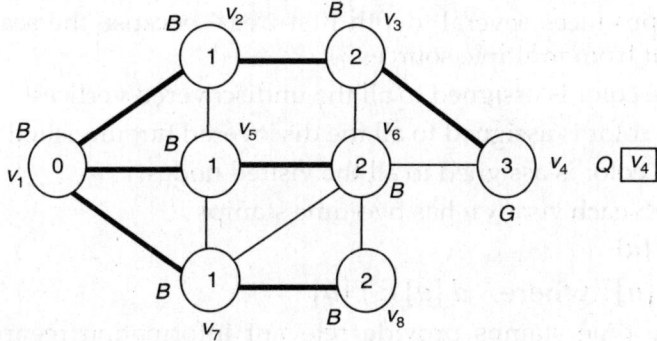

$u = v_4$

Vertices adjacent to $v_4$ are $v_3$ and $v_6$, if condition is false for both of them.

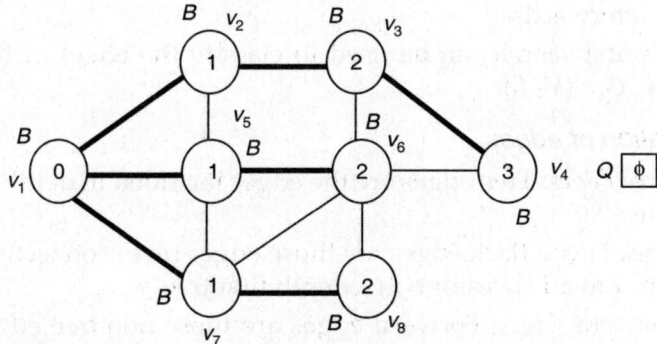

- The thick lines denote the breadth first tree at every step.

## 3.4.3 Depth First Search (DFS)

- Consider a graph $G = (V, E)$

  where, $V$ represents vertices of graph $G$

  $E$ represents edges of graph $G$

- Depth first search explores edges out of the most recently discovered vertex $v$ that still has unexplored edges leaving it.
- Once all of $v$'s edges have been explored, the search backtracks to explore edges leaving the vertex from which $v$ was discovered. This process continues until the procedure discovers all the vertices that are reachable from the original source vertex.
- If any undiscovered vertices remain, the DFS selects one of them as a new source and it repeats the search from that source.

- The DFS algorithm repeats the entire process until it has discovered every vertex.
- DFS produces several 'depth first trees' because the search may repeat from multiple sources.
- White color is assigned to all the undiscovered vertices.
- Gray color is assigned to all the discovered but unvisited nodes.
- Black color is assigned to all the visited nodes.
- In DFS each vertex $u$ has two time stamps

  (i) $d[u]$

  (ii) $f[u]$   where,   $d[u] < f[u]$

  These time stamps provide relevant information regarding the structure of the graph.

  These time stamps have values between 1 and $2|V|$.

  Vertex $u$ is white before time $d[u]$, gray between $d[u]$ and $f[u]$ and black afterwards.

- Depth first search can be used to classify the edges of the input graph $G = (V, E)$.

### Classification of edges:

  (i) *Tree Edges:* Tree edges are the edges included in depth first forest.

  (ii) *Back Edges:* Back edges are those edges $(u, v)$ connecting a vertex $u$ to an ancestor $v$ in a depth first tree.

  (iii) *Forward Edges:* Forward edges are those non tree edges $(u, v)$ connecting a vertex $u$ to a descendant $v$ in a depth first tree.

  (iv) *Cross Edges:* Cross edges are the edges between vertices in the same depth first tree as long as one vertex is not an ancestor of the other or the edges between vertices in different depth first trees.

### Algorithm

```
        DFS (G)
    Line 1      for each vertex u ∈ V [G]
    Line 2          color [u] ← white
    Line 3          π [u] ← nil
    Line 4      time ← 0
    Line 5      for each vertex u ∈ V [G]
    Line 6          if color [u] = = white
    Line 7              DFS-visit (G, u)
```

## Explanation

- The input to the DFS algorithm is a graph G which may be directed or undirected.
- Line 1 indicates the beginning of for loop that ends with Line 3. This loop is applicable for each of the vertex that belongs to the set of vertices of graph G.
- Line 2 indicates that white color is assigned to vertex $u$.
  Attribute color $[u]$ stores the color of the vertex $u$.
- Line 3 indicates that $\pi [u]$ is set to nil.
  Attribute $\pi [u]$ stores the parent of vertex $u$.
- Line 4 indicates that variable time is set to 0.
  time is a global variable which is used for time stamping.
- Line 5 indicates the beginning of for loop that ends with Line 7. This loop is applicable for each of the vertex that belongs to the set of vertices of graph G.
- Line 6 checks the *if* condition. This condition is true if the color of vertex $u$ is white.
  If this condition is true then the execution of Line 7 takes place.
- Line 7 calls the procedure DFS-visit ($G, u$).

## Algorithm

```
DFS-visit (G, u)
Line 1      time ← time + 1
Line 2      d [u] ← time
Line 3      color [u] ← gray
Line 4      for each v ∈ Adj [u]
Line 5            if color [v] = = white
Line 6                   π [v] ← u
Line 7                   DFS-visit (G, v)
Line 8      color [u] ← Black
Line 9      time ← time + 1
Line 10     f [u] ← time
```

## Explanation

- Line 1 indicates that the global variable time is incremented by 1.
- Line 2 indicates that the new value of time is recorded in the attribute $d [u]$.
  The attribute $d [u]$ records the time, when the DFS procedure discovers vertex $u$.
- Line 3 indicates that gray color is assigned to vertex $u$.

- Line 4 indicates the beginning of for loop that ends with Line 7. This loop is applicable for each vertex $v$ that belongs to the adjacency list of $u$.
- Line 5 checks the *if* condition. If this condition is true then the execution of Line 6 and Line 7 takes place. This condition is true when the color of vertex $v$ is white.
- Line 6 indicates that vertex $u$ is the parent of vertex $v$.
- Line 7 calls the procedure DFS-visit $(G, v)$.
- Line 8 indicates that black color is assigned to vertex $u$.
- Line 9 indicates that the variable time is incremented by one.
- Line 10 indicates that the finishing time is recorded in the attribute $f[u]$.

  The attribute $f[u]$ records the time when the DFS procedure finishes vertex $u$, i.e. the time when the DFS stops examining the adjacency list of vertex $u$.

## Analysis

- The loops of Lines 1-3 and Lines 5-7 take $\Theta(|V|)$ time, exclusive of the time to execute the calls to DFS-visit.
- The procedure DFS-visit is called exactly once for each vertex $v \in V$, since the vertex $u$ on which DFS-visit is invoked must be white and the first thing DFS-visit does is paint vertex $u$ gray.
- During an execution of DFS-visit $(G, v)$, the loop of Lines 4-7 executes $|\text{Adj}[v]|$ times. Since $\sum_{v \in V} \text{Adj}[v] = \Theta(|E|)$ the total cost of executing lines 4-7 of DFS-visit is $\Theta(|E|)$.
- Therefore, the running time of DFS is $\Theta(|V| + |E|)$.

## Example

Apply DFS on the following directed graph.

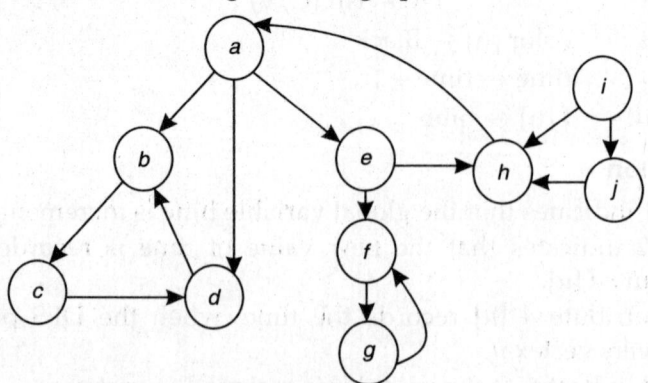

**Solution**

DFS (*G*)

Following figure is obtained after the execution of first for loop.

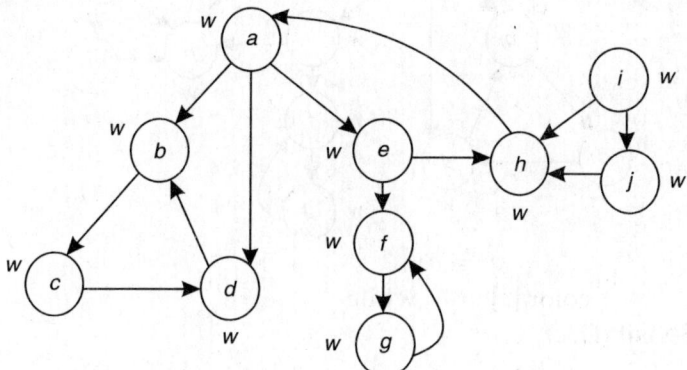

*W* represents white color

$$\text{time} = 0$$

for $\qquad u = a$

$\qquad$ color [*a*] = = white

DFS-visit (*G*, *a*)

$$\text{time} = 1$$

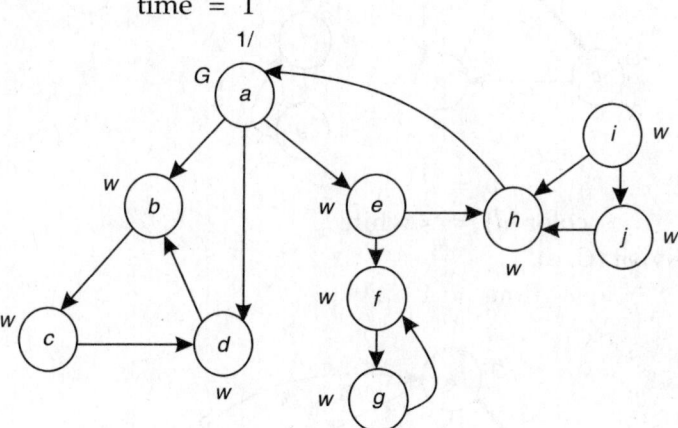

- *G* represents gray color
- The values of attributes *d* [*d*] / *f* [*u*] are appeared adjacent to the vertices
- Vertices adjacent to *a* are *b*, *d* and *e*

for $\qquad v = b$

$\qquad$ color [*b*] = = white

DFS-visit (*G*, *b*)

$$\text{time} = 2$$

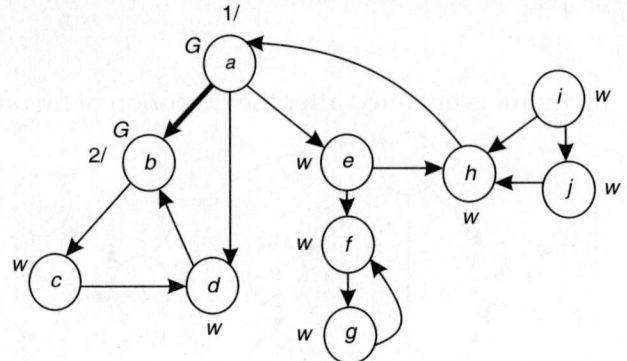

$$v = c$$
$$\text{color }[c] = = \text{white}$$
DFS-visit $(G, c)$
$$\text{time} = 3$$

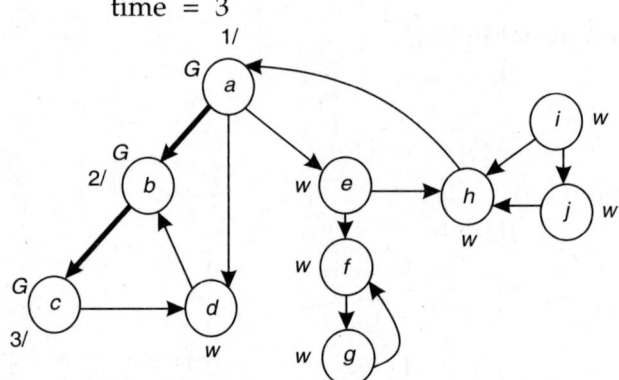

$$v = d$$
$$\text{color }[d] = = \text{white}$$
DFS-visit $(G, d)$
$$\text{time} = 4$$

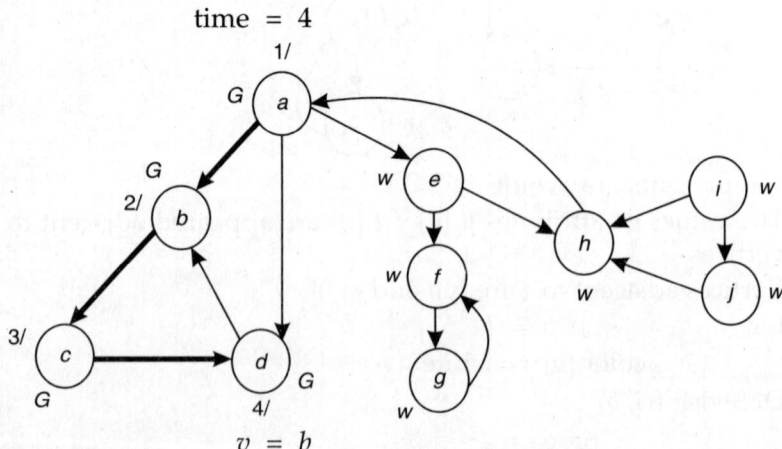

$$v = b$$

color $[b] \neq$ white

so, if condition fails

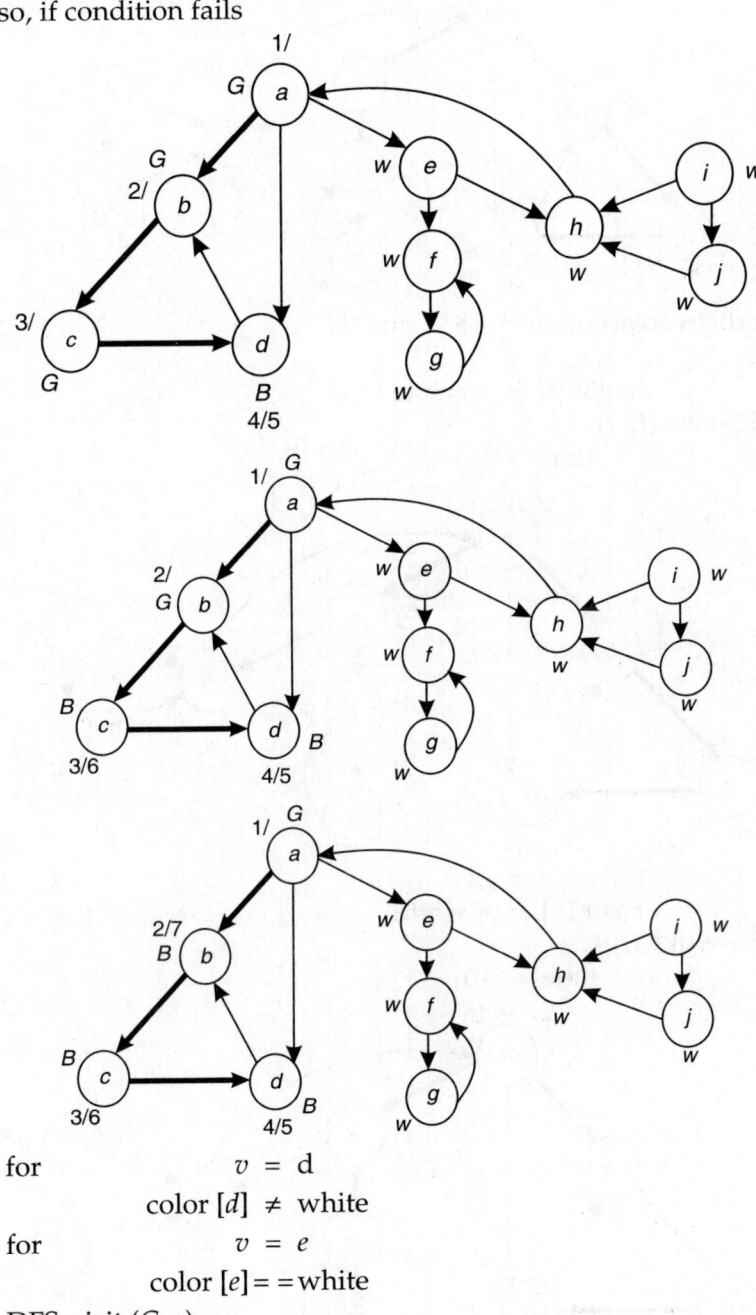

for $v = $ d

color $[d] \neq$ white

for $v = $ e

color $[e] == $ white

DFS-visit $(G, e)$

time $= 8$

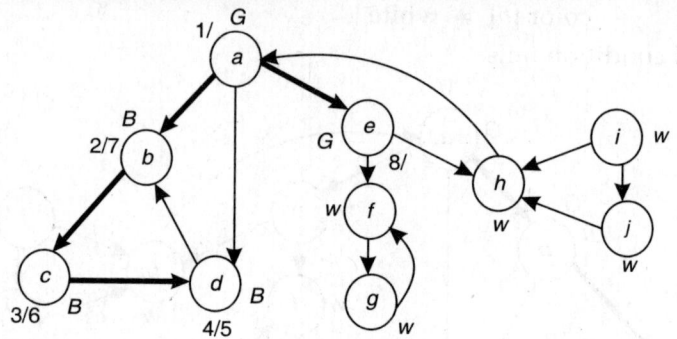

Vertices adjacent to '*e*' are '*f*' and '*h*'
for                $v = f$
            color [*f*] = = white
DFS-visit $(G, f)$
                time = 9

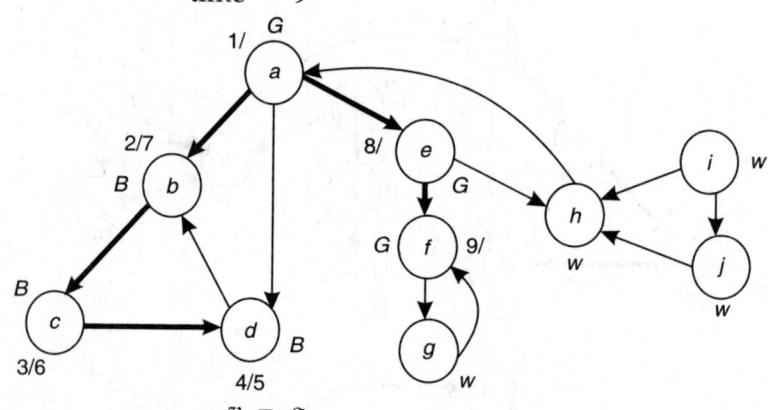

                $v = g$
            color [*g*] = = white
DFS-visit $(G, g)$
                time = 10

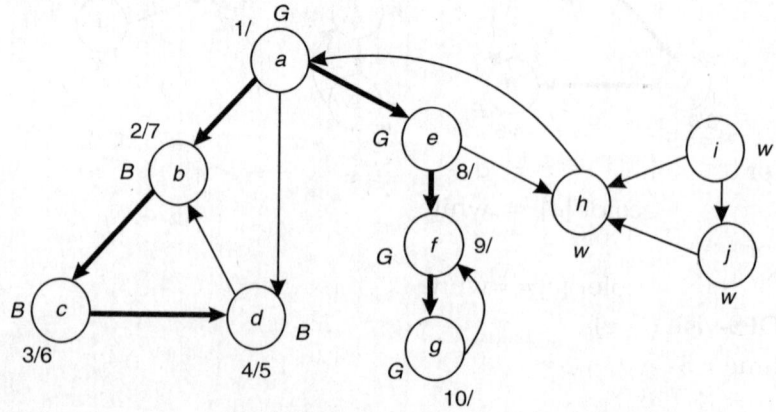

$$v = f$$
$$\text{color } [f] \neq \text{white}$$

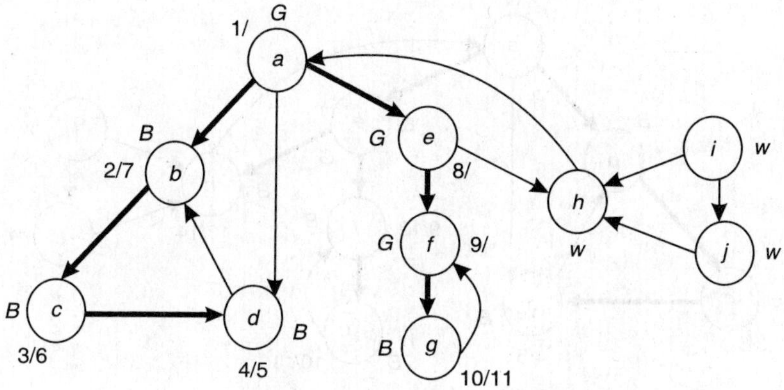

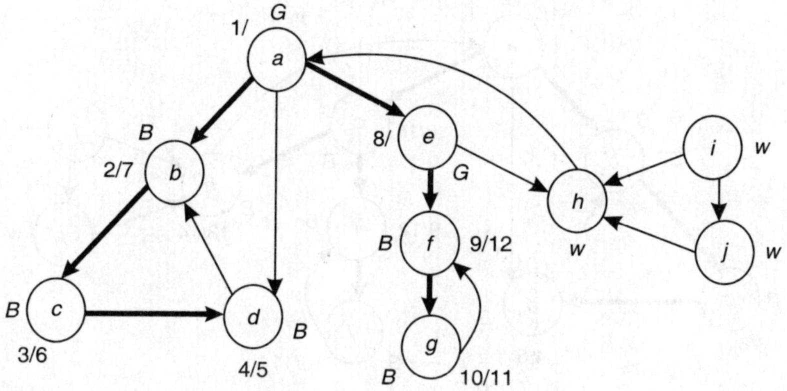

$$v = f$$
$$\text{color } [f] \neq \text{white}$$

for                                    $$v = h$$
$$\text{color } [h] = = \text{white}$$
DFS-visit $(G, h)$
$$\text{time} = 13$$

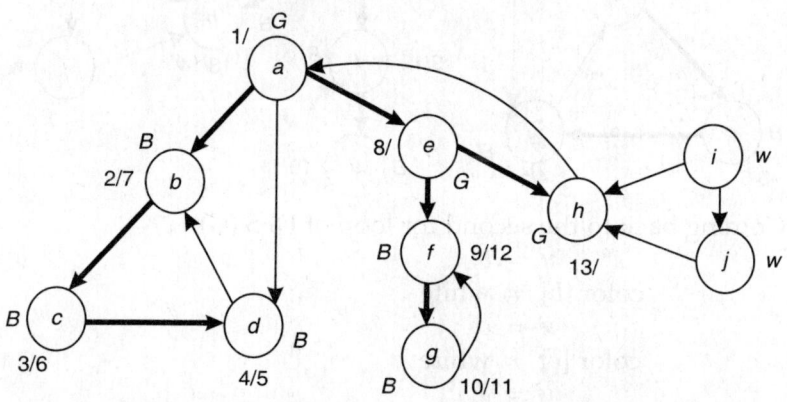

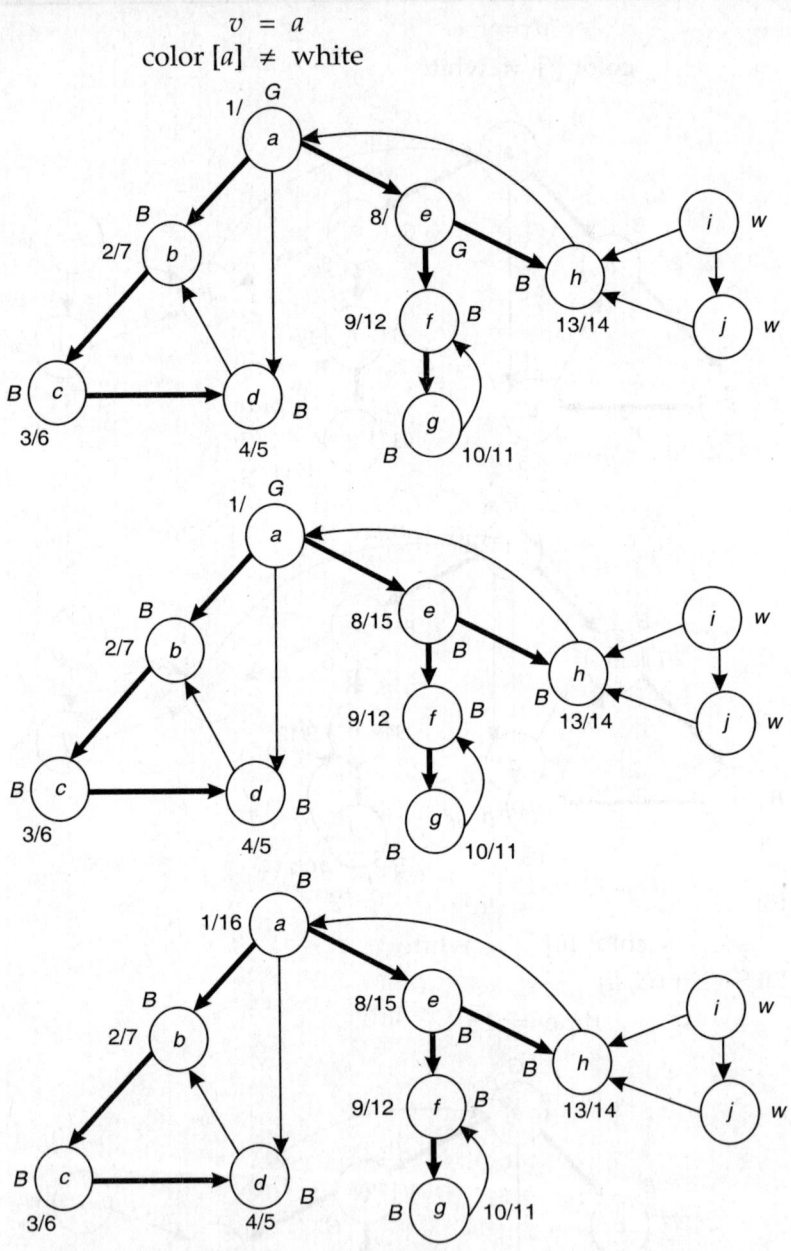

$v = a$

color $[a] \neq$ white

Coming back to the second for loop of DFS $(G)$

$$u = b$$
$$\text{color } [b] \neq \text{white}$$
$$u = c$$
$$\text{color } [c] \neq \text{white}$$
$$u = d$$

$$\text{color } [d] \neq \text{white}$$
$$u = e$$
$$\text{color } [e] \neq \text{white}$$
$$u = f$$
$$\text{color } [f] \neq \text{white}$$
$$u = g$$
$$\text{color } [g] \neq \text{white}$$
$$u = h$$
$$\text{color } [h] \neq \text{white}$$
$$u = i$$
$$\text{color } [i] = = \text{white}$$
DFS-visit $(G, i)$
$$\text{time} = 17$$

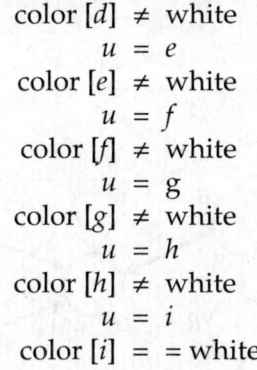

Vertices adjacent to $i$ are $h$ and $j$

for $\qquad v = h$
$$\text{color } [h] \neq \text{white}$$
for $\qquad v = j$
$$\text{color } [j] = = \text{white}$$
DFS-visit $(G, j)$
$$\text{time} = 18$$

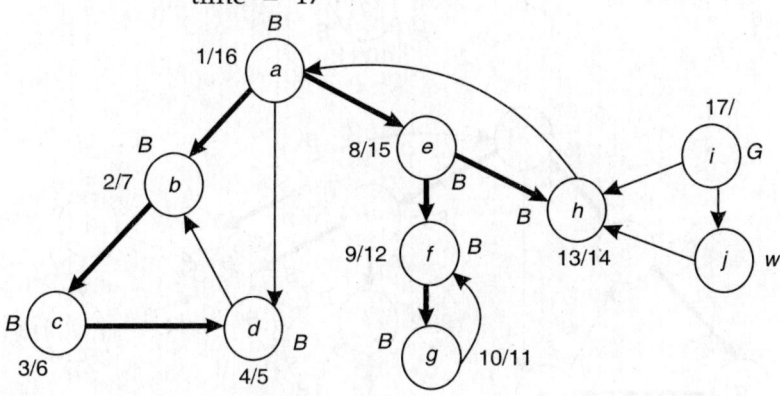

Vertex adjacent to $j$ is $h$

color $[h] \neq$ white

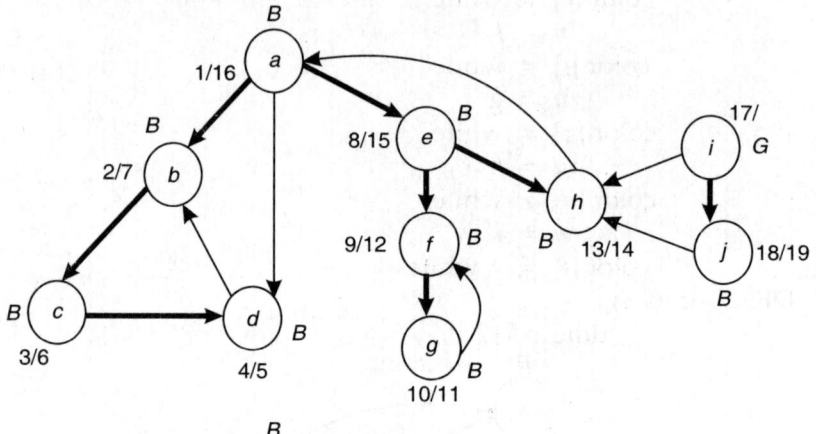

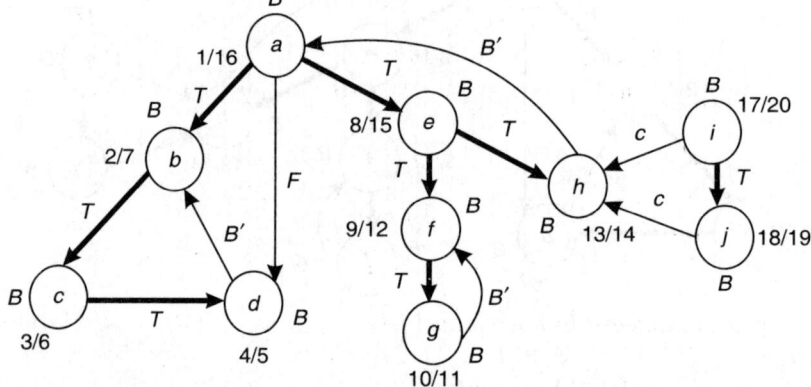

Coming back to the second for loop of DFS $(G)$

$$u = j$$

color $[j] \neq$ white

Dark lines denote depth first tree

$B', F, C$ and $T$ represent back, forward, cross and tree edges respectively.

## 3.5 DATA STRUCTURE BASED APPROACH

- Efficient data structures are the key to design efficient algorithms.
- Algorithms use data structures to manipulate the data contained in them.
- In data structure based approach a specific data structure is used to provide a solution for a certain problem.
- For example 'heap' data structure is useful for heap sort.

## 3.5.1 Heap Sort

- The heap sort uses a heap data structure to sort the elements.
- A heap data structure is an array object that can be viewed as a nearly complete binary tree. The tree is completely filled on all levels except possibly the lowest.
- An array A that represents a heap has two attributes:
  (i) length [A]: It gives the number of elements in the array.
  (ii) Heap-size [A]: It represents the number of elements in the heap that are stored within array A.

*Height of a heap*: The height of a heap is the height of its root. In general the height of a node in a heap is the number of edges on the longest simple downward path from the node to a leaf. A heap of $n$ elements is based on a complete binary tree, so, its height is $\Theta$ (lg $n$).

*Addressing in a heap*: Consider a node in a heap whose address is $i$ then,

$$\text{Address of its parent node} = \lfloor i/2 \rfloor$$
$$\text{Address of its left child} = 2i$$
$$\text{Address of its right child} = 2i + 1$$

### *Types of heap*

  (i) *Max-heap*: It satisfies the max-heap property which says that the value of a node is at most the value of its parent. So, the largest element in a max-heap is stored at the root and the subtree rooted at a node contains values no larger than that contained at the node itself.

  So, for every node i other than the root,

  A [Parent ($i$)] $\geq$ A [$i$]

  (ii) *Min-heap*: It satisfies the min-heap property which says that the value of a node is atleast the value of its parent. So, the smallest element in a min-heap is at the root.

  For every node i other than the root

  A [Parent ($i$)] $\leq$ A [$i$]

### Algorithm

Max-Heapify (A, i)

| | |
|---|---|
| Line 1 | $\ell \leftarrow$ Left (i) |
| Line 2 | r $\leftarrow$ Right (i) |
| Line 3 | if $\ell \leq$ heap-size [A] and A [$\ell$] > A [i] |
| Line 4 | largest $\leftarrow \ell$ |
| Line 5 | else   largest $\leftarrow$ i |
| Line 6 | if  r $\leq$ heap-size [A] and A [r] > A [largest] |

Line 7                    largest ← r
Line 8        if largest ≠ i
Line 9                    exchange A [i] with A [largest]
Line 10                   Max-Heapify (A, largest)

## Explanation

- This procedure is called to maintain the max-heap property.
- The inputs of this procedure are an array A and an index $i$ into the array.
- Max-Heapify assumes that the binary trees rooted at Left ($i$) and Right ($i$) are max heaps but that A [$i$] might be smaller than its children, thus violating the max-heap property.
- Max-Heapify procedure ensures that the subtree rooted at index $i$ obeys the max-heap property.
- Line 1 indicates that the index of left child of node $i$ is stored in $\ell$.
- Line 2 indicates that the index of right child of node $i$ is stored in $r$.
- Line 3 checks the *if* condition. If this condition is true then the execution of Line 4 takes place else the control goes to Line 5. The *if* condition is true when $\ell$ is less than or equal to the size of heap A and the heap element A [$\ell$] is greater than the heap element A [$i$].
- Line 4 indicates that the index of left child of node $i$ is stored in largest.
- Line 5 indicates that largest stores index $i$ .
- Line 6 checks the *if* condition. If this condition is true then the execution of Line 7 takes place else the procedure checks the *if* condition of Line 8. The *if* condition of Line 6 is true when $r$ is less than or equal to the size of heap A and the heap element A[$r$] is greater than the heap element A [largest].
- Line 7 indicates that largest stores the index of right child of node $i$.
- Line 8 checks the *if* condition. This condition is true when the index stored in largest is not equal to index $i$. If this condition is true then the execution of Line 9 and Line 10 takes place.
- Line 9 indicates that the heap element A [$i$] is exchanged with the heap element A [largest].
- Line 10 calls the procedure Max-Heapify (A, largest).
- The above procedure Max-Heapify (A, $i$) terminates, when the *if* condition of Line 8 becomes false, i.e. largest = = $i$.
- In the procedure Max-Heapify (A, $i$), the goal is to determine the largest of the elements A[$i$], A [$\ell$] and A [$r$] and then stores its index in largest.

- When A $[i]$ is largest it means largest $= = i$, which implies that the subtree rooted at node $i$ is a max-heap and the procedure terminates.
- When A $[\ell]$ or A $[r]$ is largest, then A$[i]$ is swapped with A[largest] causing node $i$ and its children to satisfy the max-heap property.
- Now, the node indexed by largest has the original value A$[i]$ and the subtree rooted at largest might violate the max-heap property. Thus, we call Max-Heapify procedure recursively on that subtree.

## Analysis

- The running time of Max-Heapify on a subtree of size $n$ rooted at a given node $i$ is the $\Theta$ (1) time to fix up the relationships among the elements. A $[i]$, A [Left $(i)$], A [Right $(i)$] plus the time to run Max-Heapify on a subtree rooted at one of the children of node $i$ (assuming that the recursive call occurs).
- The children's subtrees, each have size at most $2n/3$. The worst case occurs when the bottom level of the tree is exactly half full.Therefore, the running time of Max-Heapify can be described by the recurrence:

$$T (n) \leq T (2n/3) + \Theta (1)$$

The solution to this recurrence is:

$$T (n) = O (\lg n)$$

- Alternatively, we can characterize the running time of Max-Heapify on a node of height $h$ as $O (h)$.

## Example

Illustrate the operation of Max-Heapify $(A, 3)$ on the following array:

$$A = \boxed{27}\ \boxed{17}\ \boxed{3}\ \boxed{16}\ \boxed{13}\ \boxed{10}\ \boxed{1}\ \boxed{5}\ \boxed{7}\ \boxed{12}\ \boxed{4}\ \boxed{8}\ \boxed{9}\ \boxed{0}$$

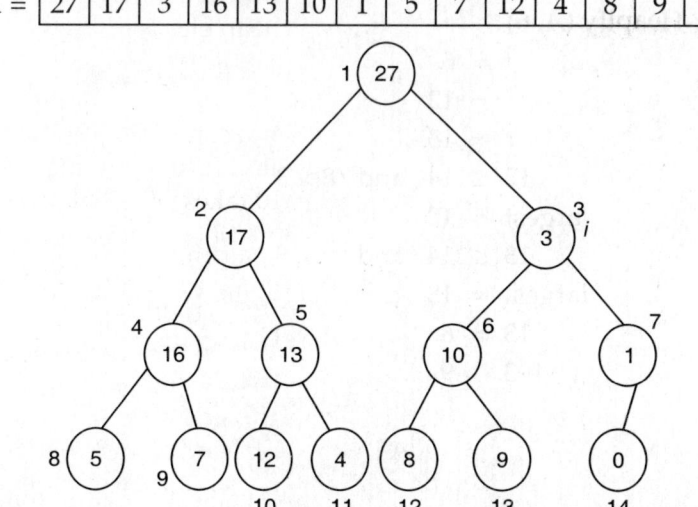

Heap-size [A] = 14

$$i = 3$$

Max-Heapify (A, 3)

$$\ell = 6$$

$$r = 7$$

$$6 \leq 14 \text{ and } 10 > 3$$

$$\text{largest} = 6$$

$$7 \leq 14 \text{ and } 1 \not> 10$$

$$3 \neq 6$$

$$3 \leftrightarrow 10$$

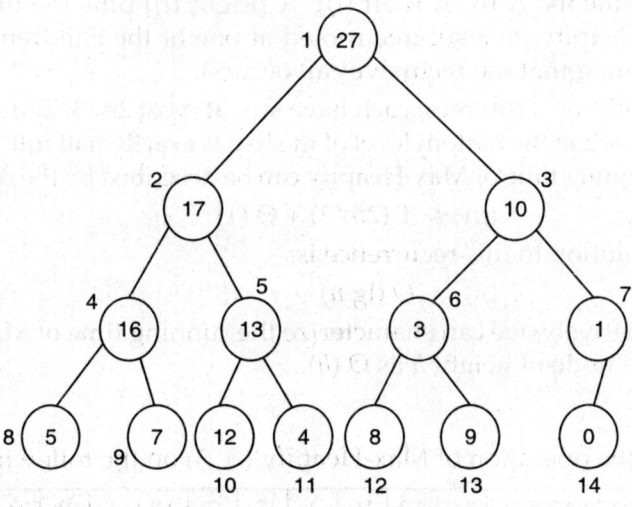

Max-Heapify (A, 6)

$$i = 6$$

$$\ell = 12$$

$$r = 13$$

$$12 \leq 14 \quad \text{and} \quad 8 > 3$$

$$\text{largest} = 12$$

$$13 \leq 14 \quad \text{and} \quad 9 > 8$$

$$\text{largest} = 13$$

$$13 \neq 6$$

$$3 \leftrightarrow 9$$

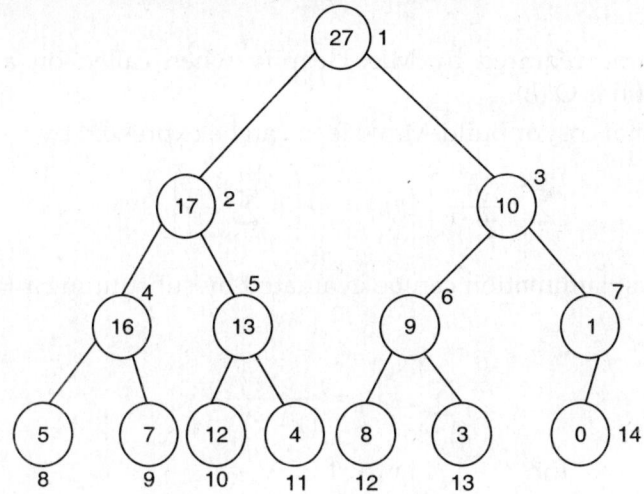

Max-Heapify $(A, 13)$

$$i = 13$$
$$\ell = 26$$
$$r = 27$$
$$26 \nleq 14$$
$$\text{largest} = 13$$
$$27 \nleq 14$$
$$13 = = 13$$

So, procedure terminates.

## Algorithm

    Build-Max-Heap (A)
    Line 1        Heap-size [A] ← length [A]

Line 2        for   $i \leftarrow \left\lfloor \dfrac{\text{length } [A]}{2} \right\rfloor$ down to 1.

Line 3                Max-Heapify (A, $i$)

## Explanation

- Line 1 indicates that the size of the heap $A$ will be equal to the length of the array $A$.
- Line 2 indicates the beginning of for loop that ends with Line 3.

   This loop is applicable for $i$ equals $\left\lfloor \dfrac{\text{length } [A]}{2} \right\rfloor$ down to 1.

- Line 3 calls the procedure Max-Heapify (A, $i$).
- The above Build-Max-Heap ($A$) procedure produces a max-heap from an unordered input array.

## Analysis

- The time required by Max-Heapify when called on a node of height h is $O(h)$.

  The total cost of Build-Max-Heap can be expressed by

  $$\sum_{h=0}^{\lfloor \lg n \rfloor} \left\lceil \frac{n}{2^{h+1}} \right\rceil O(h) = O\left( n \sum_{h=0}^{\lfloor \lg n \rfloor} \frac{h}{2^h} \right)$$

  The last summation can be evaluated by substituting $x = \frac{1}{2}$ in the formula

  $$\sum_{k=0}^{\infty} kx^k = \frac{x}{(1-x)^2}$$

  for $\qquad |x| < 1$

  that yields

  $$\sum_{h=0}^{\infty} \frac{h}{2^h} = \frac{1/2}{(1-1/2)^2}$$

  $$= 2$$

  Thus, we can bind the running time of Build-Max-Heap as

  $$O\left( n \sum_{h=0}^{\lfloor \lg n \rfloor} \frac{h}{2^h} \right) = O\left( n \sum_{h=0}^{\infty} \frac{h}{2^h} \right)$$

  $$= O(n)$$

## Example

Illustrate the operation of Build-Max-Heap (A) on the given array:

$A = $ | 4 | 5 | 8 | 7 | 3 | 9 | 10 | 1 | 11 | 6 |

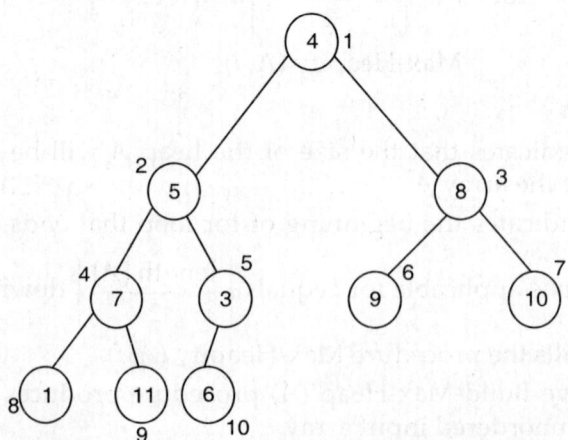

Heap-size [A] = 10

for loop is applicable for $i = \left\lfloor \dfrac{10}{2} \right\rfloor$ down to 1 = 5 down to 1

for                    $i = 5$

Max-Heapify (A, 5)

$$\ell \;=\; 10$$

$$r \;=\; 11$$

$$10 \;\le\; 10 \text{ and } 6 > 3$$

largest $= 10$

$$11 \;\nleq\; 10$$

$$10 \;\ne\; 5$$

$$3 \leftrightarrow 6$$

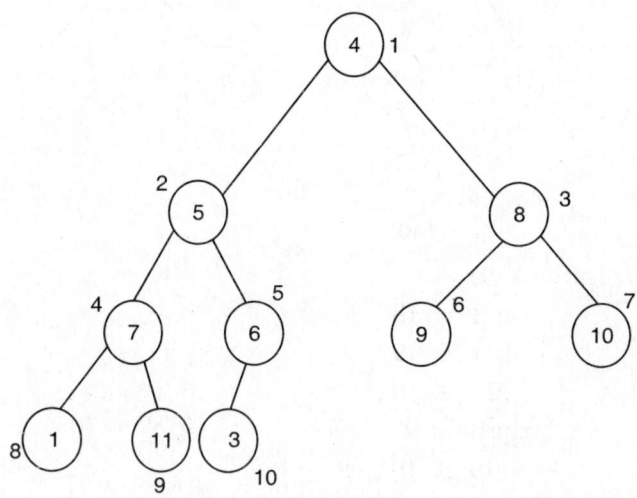

Max-Heapify (A, 10)

$$\ell \;=\; 20$$

$$r \;=\; 21$$

$$20 \;\nleq\; 10$$

largest $= 10$

$$21 \;\nleq\; 10$$

$$10 \;= \;= 10$$

So, Max-Heapify procedure terminates.

for                    $i = 4$

Max-Heapify (A, 4)

$$\ell \;=\; 8$$

$$r \;=\; 9$$

$$8 \;\le\; 10 \text{ and } 1 \ngtr 7$$

$$\text{largest} = 4$$
$$9 \le 10 \text{ and } 11 > 7$$
$$\text{largest} = 9$$
$$9 \ne 4$$
$$7 \leftrightarrow 11$$

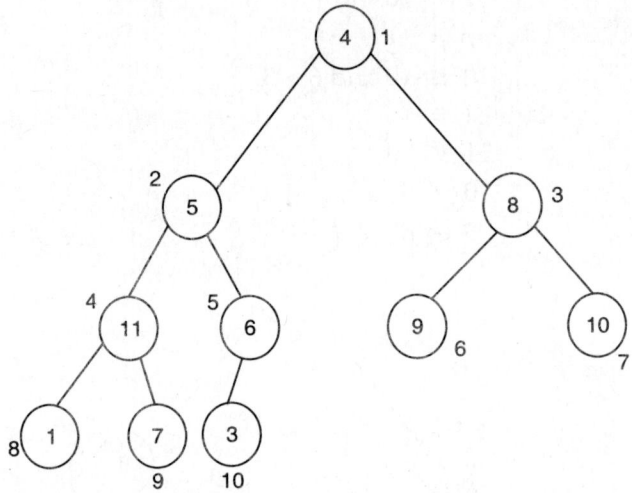

Max-Heapify $(A, 9)$
$$\ell = 18$$
$$r = 19$$
$$18 \nleq 10$$
$$\text{largest} = 9$$
$$19 \nleq 10$$
$$9 = = 9$$

Max-Heapify procedure terminates.

for $\qquad i = 3$

Max-Heapify $(A, 3)$
$$\ell = 6$$
$$r = 7$$
$$6 \le 10 \text{ and } 9 > 8$$
$$\text{largest} = 6$$
$$7 \le 10 \text{ and } 10 > 9$$
$$\text{largest} = 7$$
$$7 \ne 3$$
$$8 \leftrightarrow 10$$

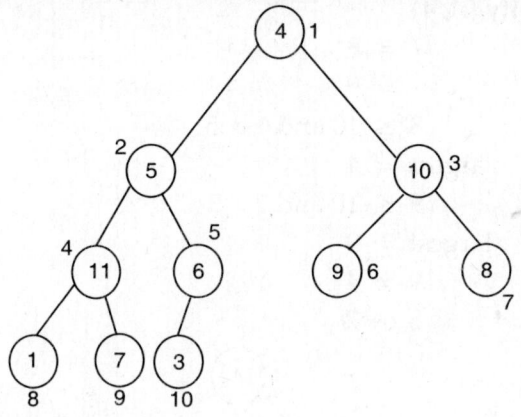

Max-Heapify $(A, 7)$

$$\ell = 14$$
$$r = 15$$
$$14 \nleq 10$$
$$\text{largest} = 7$$
$$15 \nleq 10$$
$$7 = = 7$$

Max-Heapify procedure terminates.

for                 $i = 2$

Max-Heapify $(A, 2)$

$$\ell = 4$$
$$r = 5$$
$$4 \leq 10 \text{ and } 11 > 5$$
$$\text{largest} = 4$$
$$5 \leq 10 \text{ and } 6 \ngtr 11$$
$$4 \neq 2$$
$$5 \leftrightarrow 11$$

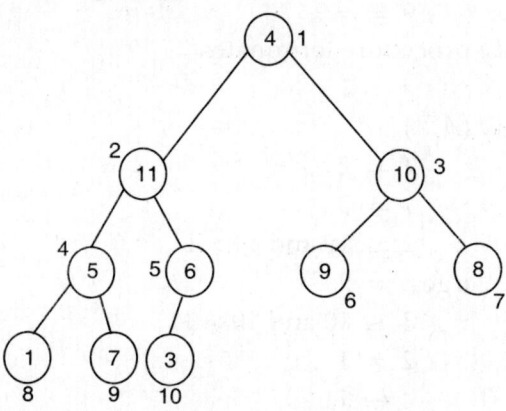

Max-Heapify $(A, 4)$

$$\ell = 8$$
$$r = 9$$
$$8 \leq 10 \text{ and } 1 \not> 5$$
$$\text{largest} = 4$$
$$9 \leq 10 \text{ and } 7 > 5$$
$$\text{largest} = 9$$
$$9 \neq 4$$
$$5 \leftrightarrow 7$$

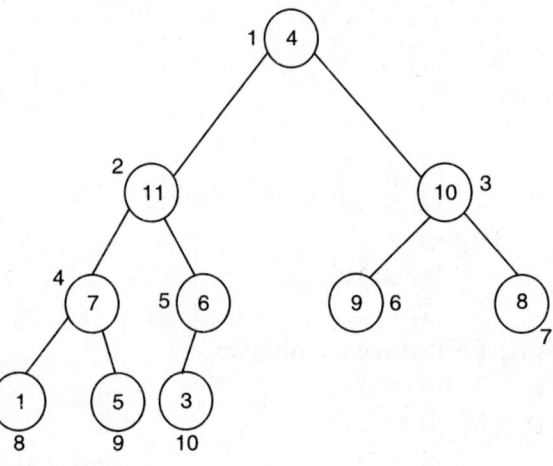

Max-Heapify $(A, 9)$

$$\ell = 18$$
$$r = 19$$
$$18 \nleq 10$$
$$\text{largest} = 9$$
$$19 \nleq 10$$
$$9 = = 9$$

Max-Heapify procedure terminates

for                    $i = 1$

Max-Heapify $(A, 1)$

$$\ell = 2$$
$$r = 3$$
$$2 \leq 10 \text{ and } 11 > 4$$
$$\text{largest} = 2$$
$$3 \leq 10 \text{ and } 10 \not> 11$$
$$2 \neq 1$$
$$4 \leftrightarrow 11$$

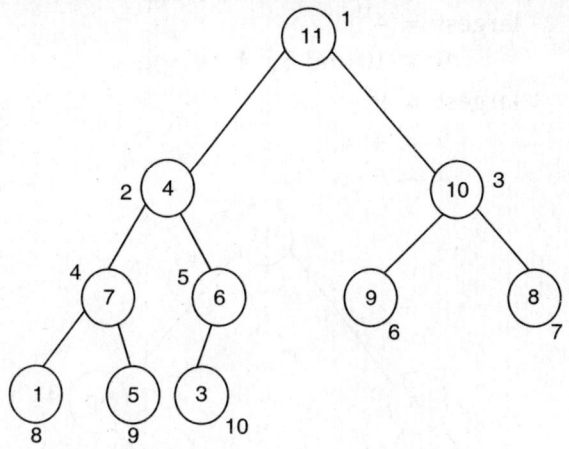

Max-Heapify $(A, 2)$

$$\ell = 4$$
$$r = 5$$
$$4 \le 10 \text{ and } 7 > 4$$
largest $= 4$
$$5 \le 10 \text{ and } 6 \ngtr 7$$
$$4 \ne 2$$
$$4 \leftrightarrow 7$$

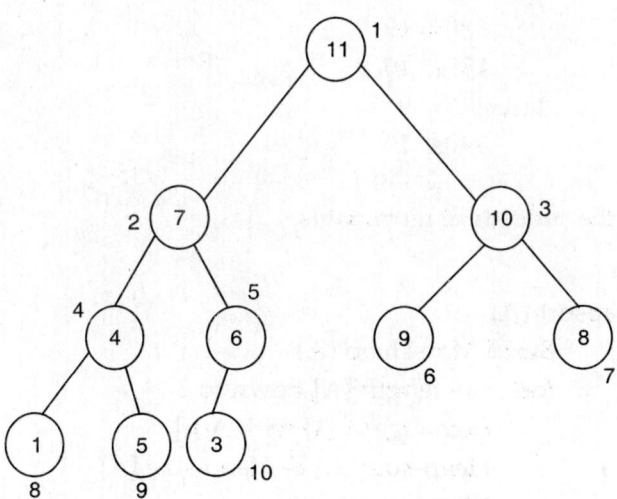

Max-Heapify $(A, 4)$

$$\ell = 8$$
$$r = 9$$
$$8 \le 10 \text{ and } 1 \ngtr 4$$

$$largest = 4$$
$$9 \leq 10 \text{ and } 5 > 4$$
$$largest = 9$$
$$9 \neq 4$$
$$4 \leftrightarrow 5$$

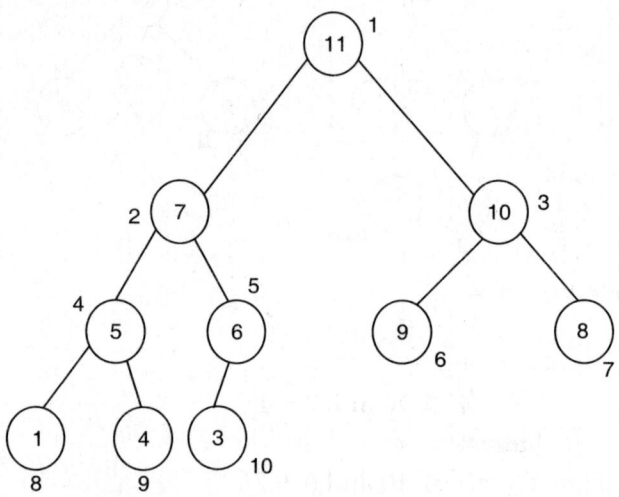

Max-Heapify (A, 9)
$$\ell = 18$$
$$r = 19$$
$$18 \nleq 10$$
$$largest = 9$$
$$19 \nleq 10$$
$$9 = = 9$$

Hence the procedure terminates.

### Algorithm

Heapsort (A)

| | |
|---|---|
| Line 1 | Build-Max-Heap (A) |
| Line 2 | for   i ← length [A] down to 2 |
| Line 3 | exchange A [1] with A [i] |
| Line 4 | Heap-size [A] ← Heap-size [A] – 1 |
| Line 5 | Max-Heapify (A, 1) |

### Explanation

- Line 1 calls the Build-Max-Heap (*A*) procedure to build a max-heap of the same size as that of the input array *A*.
- Line 2 indicates the beginning of for loop that ends with Line 5.

- Line 3 indicates that the heap element $A$ [1] is exchanged with heap element $A[i]$.
- Line 4 indicates that the size of the heap ($A$) is decremented by 1.
- Line 5 calls the procedure Max-Heapify ($A$,1).
- The above Heap sort ($A$) procedure sorts an array in place.

## Analysis

- The Heapsort procedure takes $O$ ($n$ lg $n$) time because the call to Build-Max-Heap takes $O$ ($n$) time and each of the $n - 1$ calls to Max-Heapify takes $O$ (lg $n$) time.

## Example

Illustrate the operation of Heapsort ($A$) on the array:

$$A = \boxed{4 \mid 6 \mid 5. \mid 9}$$

## Solution

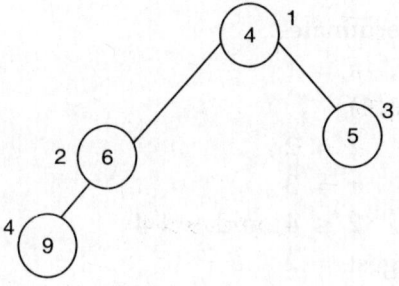

Build-Max-Heap ($A$)

Heap-size [$A$] = 4

for $\qquad i = \left\lfloor \dfrac{4}{2} \right\rfloor$ down to 1

$\qquad\qquad\quad = 2$ down to 1

for $\qquad i = 2$

Max-Heapify (A, 2)

$\qquad\qquad \ell = 4$

$\qquad\qquad r = 5$

$\qquad\qquad 4 \leq 4$ and $9 > 6$

$\qquad largest = 4$

$\qquad\qquad 5 \nleq 4$

$\qquad\qquad 4 \neq 2$

$\qquad\qquad 6 \leftrightarrow 9$

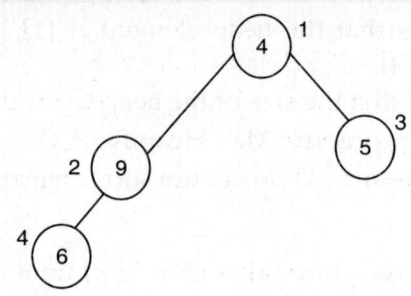

Max-Heapify $(A, 4)$

$$\ell = 8$$
$$r = 9$$
$$8 \nleq 4$$
$$largest = 4$$
$$9 \nleq 4$$
$$4 = = 4$$

So, procedure terminates.

for                     $i = 1$

Max-Heapify $(A, 1)$

$$\ell = 2$$
$$r = 3$$
$$2 \leq 4 \text{ and } 9 > 4$$
$$largest = 2$$
$$3 \leq 4 \text{ and } 5 \ngtr 9$$
$$2 \neq 1$$
$$4 \leftrightarrow 9$$

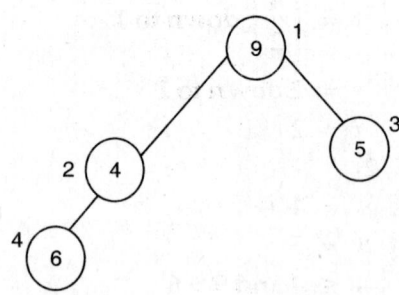

Max-Heapify $(A, 2)$

$$\ell = 4$$
$$r = 5$$
$$4 \leq 4 \text{ and } 6 > 4$$

$$\text{largest} = 4$$
$$5 \nleq 4$$
$$4 \neq 2$$
$$4 \leftrightarrow 6$$

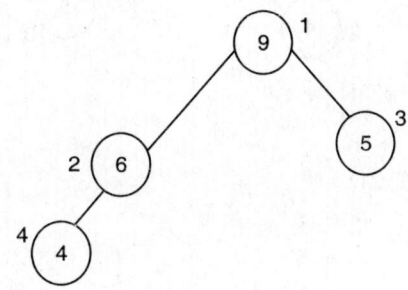

Max-Heapify $(A, 4)$

$$\ell = 8$$
$$r = 9$$
$$8 \nleq 4$$
$$\text{largest} = 4$$
$$9 \nleq 4$$
$$4 = = 4$$

So, procedure terminates.

for              $i = 4$ down to 2
for              $i = 4$
$$9 \leftrightarrow 4$$

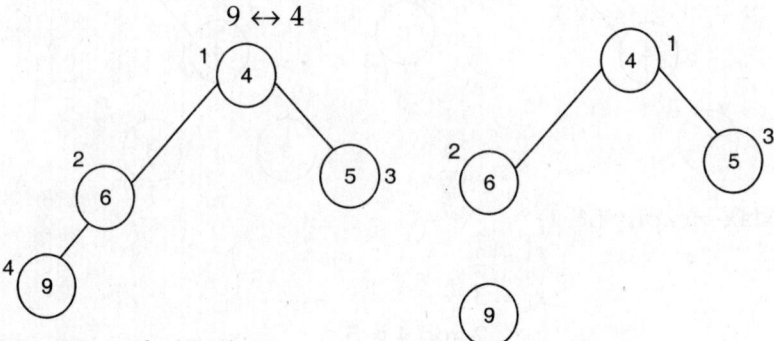

Max-Heapify $(A, 1)$

$$\ell = 2$$
$$r = 3$$
$$2 \leq 3 \text{ and } 6 > 4$$
$$\text{largest} = 2$$
$$3 \leq 3 \text{ and } 5 \ngtr 6$$
$$2 \neq 1$$
$$4 \leftrightarrow 6$$

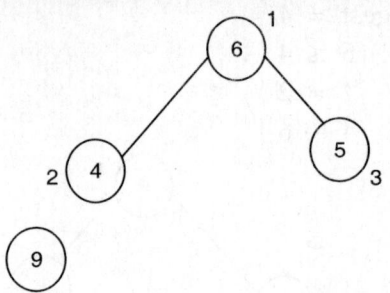

Max-Heapify $(A, 2)$

$$\ell = 4$$
$$r = 5$$
$$4 \not\leq 3$$
$$\text{largest} = 2$$
$$5 \not\leq 3$$
$$2 = \,= 2$$

So, Max-Heapify procedure terminates.

for                $i = 3$
                 $6 \leftrightarrow 5$

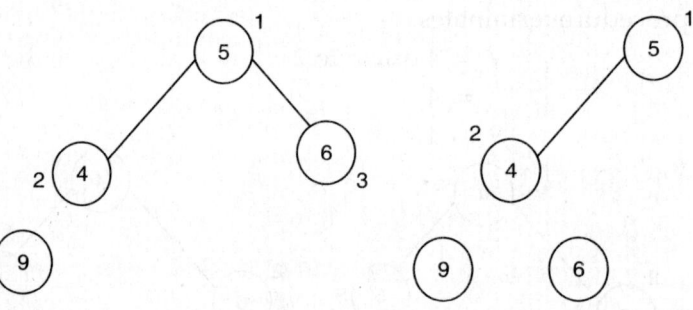

Max-Heapify $(A, 1)$

$$\ell = 2$$
$$r = 3$$
$$2 \leq 2 \text{ and } 4 \not> 5$$
$$\text{largest} = 1$$
$$3 \not\leq 2$$
$$1 = \,= 1$$

So, Max-Heapify procedure terminates.

for                $i = 2$
                 $5 \leftrightarrow 4$

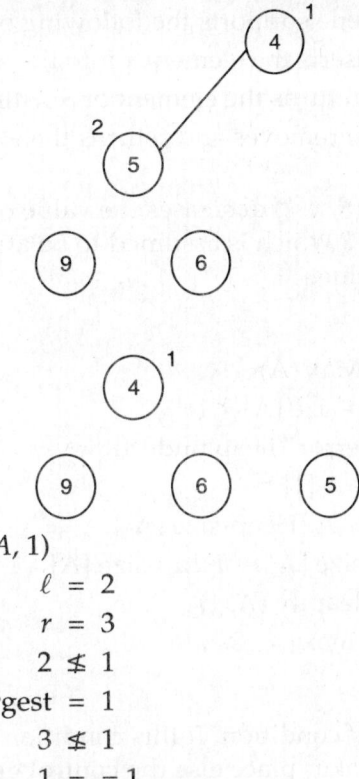

Max-Heapify $(A, 1)$

$$\ell = 2$$
$$r = 3$$
$$2 \not\leq 1$$
$$\text{largest} = 1$$
$$3 \not\leq 1$$
$$1 = \, = 1$$

So, Max-Heapify procedure terminates.

Resulting array,

$$A = \boxed{\begin{array}{|c|c|c|c|} 9 & 6 & 5 & 4 \end{array}}$$

### Priority Queues

A priority queue is a data structure that uses a heap to maintain a set of elements.

Types of priority queues:
 (i) Max-priority queues: They use max-heaps
 (ii) Min-priority queues: They use min-heaps

A max-priority queue supports the following operations:

 (a) Insert $(S, x)$, inserts the element $x$ into the set of elements $S$.

 (b) Maximum $(S)$ returns the element of $S$ with the largest key.

 (c) Extract-Max $(S)$ removes and returns the element of $S$ with the largest key.

 (d) Increase-key $(S, x, k)$ increases the value of element $x$'s key to the new value $k$ which is assumed to be at least as large as $x$'s current key value.

A min-priority queue supports the following operations:

(a) Insert $(S, x)$, inserts the element $x$ into the set of elements $S$.

(b) Minimum $(S)$ returns the element of $S$ with the smallest key.

(c) Extract-Min $(S)$ removes and returns the element of $S$ with the smallest key.

(d) Decrease-key $(S, x, k)$ decreases the value of element $x$'s key to the new value $k$ which is assumed to be at least as small as $x$'s current key value.

## Algorithm

Heap-Extract-Max (A)

Line 1        if Heap-size [A] < 1

Line 2             error "heap underflow"

Line 3        max ← A [1]

Line 4        A [1] ← A [Heap-size [A]]

Line 5        Heap-size [A] ← Heap-size [A] – 1

Line 6        Max-Heapify (A, 1)

Line 7        return max

## Explanation

- Line 1 checks the *if* condition. If this condition is true then the execution of Line 2 takes place else the control goes to Line 3. The *if* condition is true when the size of heap is less than 1.
- Line 2 shows an error and the procedure terminates.
- Line 3 indicates that max stores the heap element $A$ [1].
- Line 4 indicates that the heap element $A$ [1] is exchanged with the heap element $A$ [Heap-size [$A$]].
- Line 5 indicates that the heapsize is decremented by 1 because the new heap element $A$ [Heap-size [$A$]] having the largest key value is removed from the heap.
- Line 6 calls the procedure Max-Heapify $(A,1)$.
- Line 7 returns the max.

## Analysis

- The running time of Heap-Extract-Max is $O$ (lg $n$) because it performs a constant amount of work on top of the $O$ (lg $n$) time for Max-Heapify.

## Example

Illustrate the operation of Heap-Extract-Max $(A)$ on the following heap:

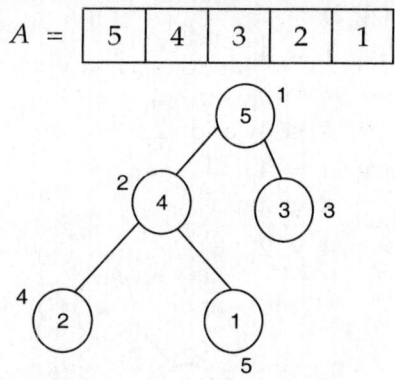

$A =$ | 5 | 4 | 3 | 2 | 1 |

Heap-size $[A] = 5$

$$5 \nless 1$$

So, if condition fails

$$\text{max} = 5$$

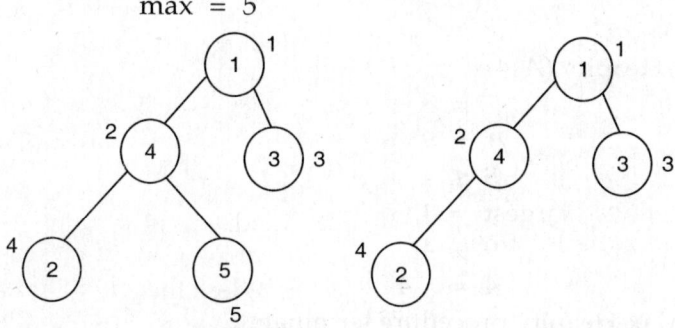

Now, Heap-size $[A] = 4$

Max-Heapify $(A, 1)$

$$\ell = 2$$
$$r = 3$$
$$2 \le 4 \text{ and } 4 > 1$$
$$\text{largest} = 2$$
$$3 \le 4 \text{ and } 3 \ngtr 4$$
$$2 \ne 1$$

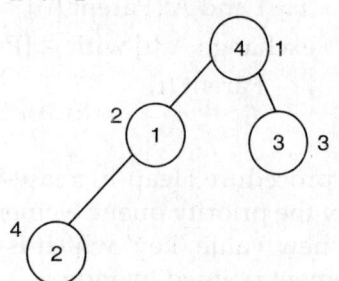

Max-Heapify $(A, 2)$

$$\ell = 4$$
$$r = 5$$
$$4 \le 4 \quad \text{and} \quad 2 > 1$$
$$\text{largest} = 4$$
$$5 \nleq 4$$
$$4 \ne 2$$

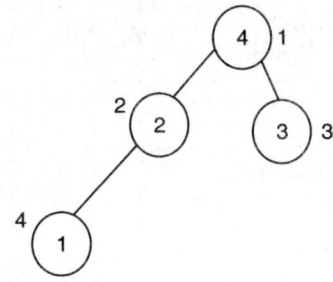

Max-Heapify $(A, 4)$

$$\ell = 8$$
$$r = 9$$
$$8 \nleq 4$$
$$\text{largest} = 4$$
$$9 \nleq 4$$
$$4 = = 4$$

So, Max-Heapify procedure terminates.

The procedure Extract-Max $(A)$ returns a heap element with key value 5.

**Algorithm**

Heap-Increase-key (A, i, key)

| | |
|---|---|
| Line 1 | if  key < A[i] |
| Line 2 | error "new key is smaller than current key" |
| Line 3 | A [i] ← key |
| Line 4 | while   i > 1 and A [Parent (i)] < A [i] |
| Line 5 | exchange A [i] with A [Parent (i)] |
| Line 6 | i ← Parent (i) |

**Explanation**

- The inputs to the procedure Heap-Increase-key are a heap '$A$', an index '$i$' to identify the priority queue element whose key we wish to increase and a new value 'key' which is to be assigned to the priority queue element pointed by index $i$.

- Line 1 checks the *if* condition. If this condition is true then the execution of Line 2 takes place else the control goes to Line 3. The *if* condition is true when the new key is smaller than the current key.
- Line 2 indicates an error and the procedure terminates.
- Line 3 indicates that the key of priority queue element $A[i]$ is updated by a new value.
- Line 4 indicates the beginning of while loop that ends with Line 6. This loop is applicable for $i$ greater than 1 and the priority queue element $A[i]$ is greater than its parent.
- Line 5 exchanges the priority queue element $A[i]$ with its parent.
- Line 6 indicates that the index to $A[i]$ now points to its parent.
- During the execution of while loop the procedure repeatedly compares a priority queue element to its parent, exchanging their keys and continuing if the element's key is larger. If the element's key is smaller than its parent, it means that the max-heap property holds and the procedure terminates.

**Analysis**

- The running time of Heap-Increase-key on an n-element heap is $O(\lg n)$ because the path traced from the node updated in Line 3 to the root has length $O(\lg n)$.

**Example**

Illustrate the operation of Heap-Increase-key $(A, 5, 7)$ on the following max-heap.

$$A = \boxed{8 \quad 6 \quad 4 \quad 5 \quad 3 \quad 2}$$

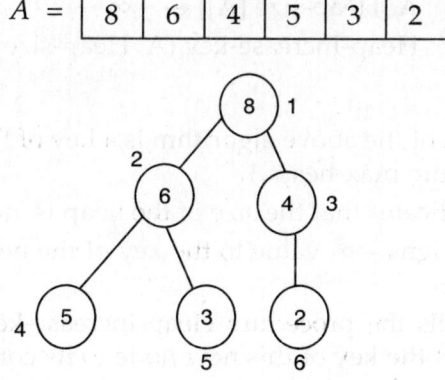

Heap-Increase-key $(A, 5, 7)$

$7 \not< 3$

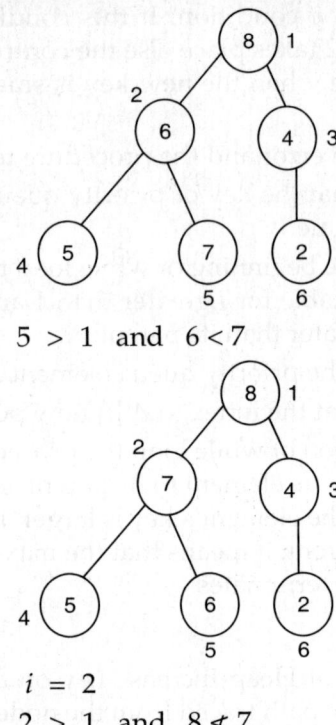

$$5 > 1 \quad \text{and} \quad 6 < 7$$

$$i = 2$$
$$2 > 1 \quad \text{and} \quad 8 \not< 7$$

So, while loop terminates

## Algorithm

Max-Heap-Insert (A, key)

Line 1        Heap-size [A] ← Heap-size [A] + 1

Line 2        A [Heap-size [A]] ← – ∞

Line 3        Heap-Increase-key (A, Heap-size [A], key)

## Explanation

- The input of the above algorithm is a key of the new element to be inserted into max-heap *A*.
- Line 1 indicates that the size of the heap is incremented by one.
- Line 2 assigns – ∞ value to the key of the new node added in the heap.
- Line 3 calls the procedure Heap-Increase-key (*A*, Heap-size [*A*], key) to set the key of this new node to its correct value and maintain the max-heap property.

## Analysis

- The running time of Max-Heap-Insert on an *n*-element heap is $O(\lg n)$.

## Example

Illustrate the operation of Max-Heap-Insert $(A, 10)$ on the following max-heap:

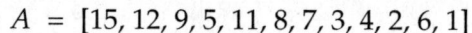

$$A = [15, 12, 9, 5, 11, 8, 7, 3, 4, 2, 6, 1]$$

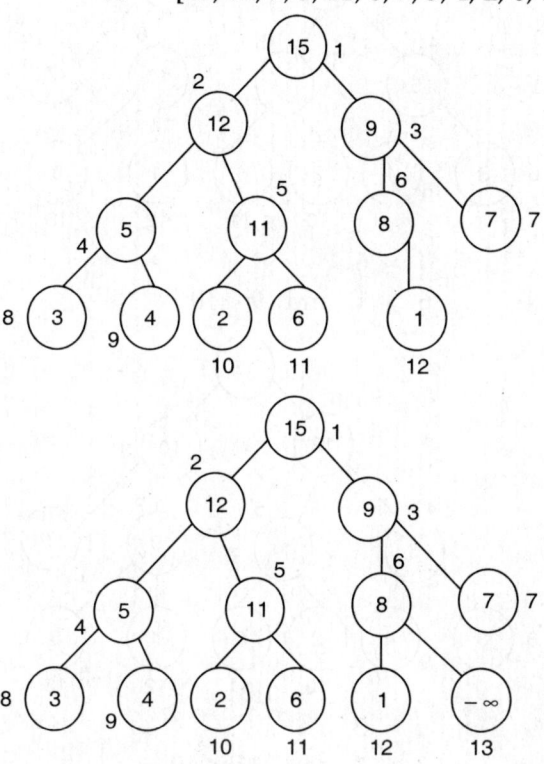

Heap-Increase-key $(A, 13, 10)$

$$10 \nless -\infty$$

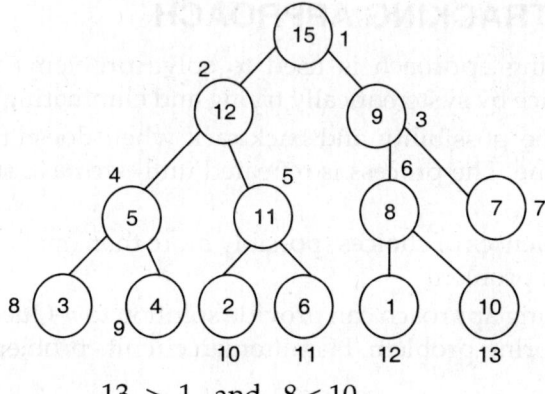

$$13 > 1 \quad \text{and} \quad 8 < 10$$

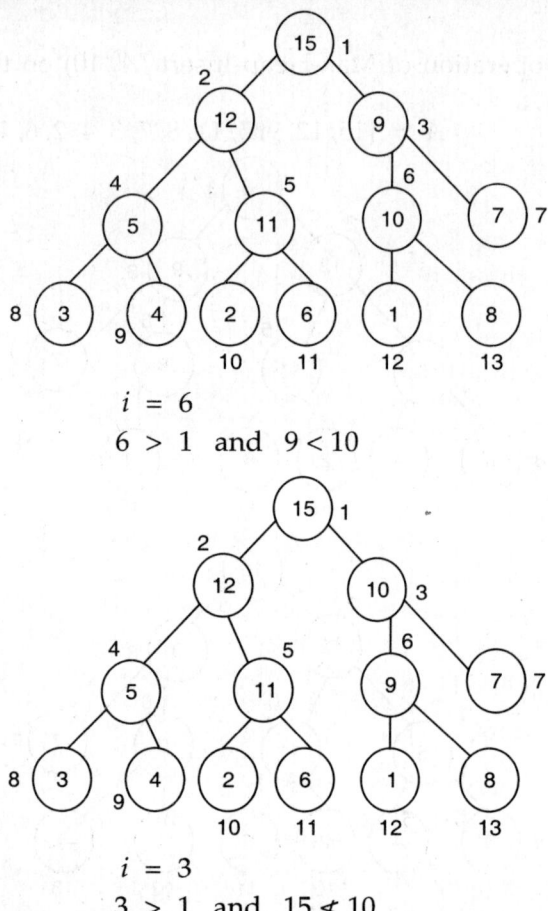

$$i = 6$$
$$6 > 1 \quad \text{and} \quad 9 < 10$$

$$i = 3$$
$$3 > 1 \quad \text{and} \quad 15 \not< 10$$

So, while loop terminates.

## 3.6 BACKTRACKING APPROACH

- Backtracking approach is used to solve problems with a large search space by systematically trying and eliminating possibilities.
- Explore the possibility and backtrack when doesn't work to try the new one. The process is repeated until we have tried all set of choices.
- Some sequences of choices (possibly more than one) may be a solution to the problem.
- Backtracking approach can provide solution to $n$-Queens problem, graph coloring problem, Hamiltonian circuits problem etc.

### 3.6.1 *n*-Queens Problem

- In *n*-Queens problem, we have to find a placement for *n*-Queens in a chess board such that neither one falls under threat of capture. It means no two queens share the same row, column or diagonal.
- Backtracking approach can provide a solution to *n*-Queens problem.
- A backtracking algorithm places one queen at a time.
- At each level of search, there are *n* possibilities, out of which we try one and move to the next level.
- If at some level we find that all *n* possibilities fail, we backtrack, i.e. go back to the previous level.
- At the previous level, we then move to the next possibility and keep trying.

**Example**

Give solutions to 4-Queens problem.

**Solution**

Following state space tree is used to find the solutions to 4-Queens problem.

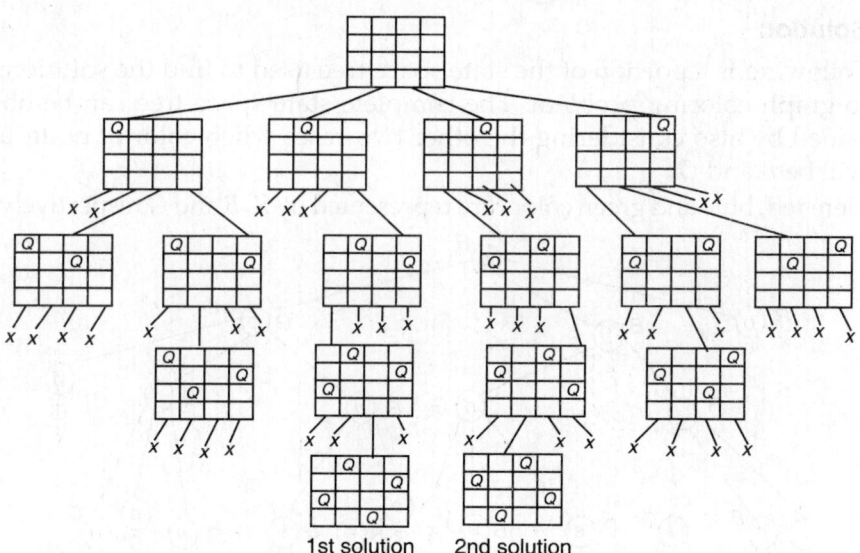

1st solution    2nd solution

- Here X denotes dead end.

### 3.6.2 Graph Coloring Problem

- Let *G* be a graph and *m* be any given positive integer, we want to discover if the nodes of *G* can be colored in such a way that no two adjacent nodes have the same color yet only *m* colors are used.

- Backtracking approach can be used to solve the graph colouring problem.
- State space tree:
    (i) Start by assigning an arbitrary color to one of the vertices.
    (ii) Continue coloring while maintaining the constraints imposed by the edges.
    (iii) If we reach a vertex that cannot be colored, backtrack, i.e. go back up the recursion tree and explore other children.

## Example

Find 3-coloring schemes for the given graph:

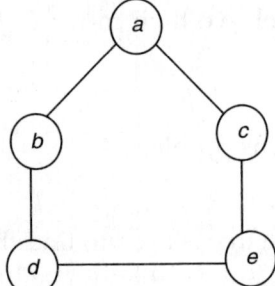

## Solution

Following is a portion of the state space tree used to find the solutions to graph coloring problem. The complete state space tree can be obtained by also considering the other two cases when color of node 'a' will be $B$ and $G$.

Here red, blue and green colors are represented by $R$, $B$ and $G$ respectively.

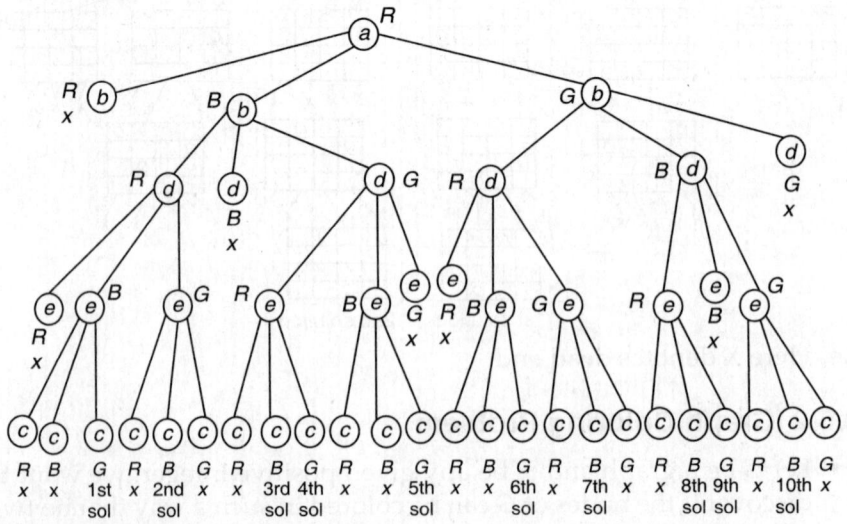

- Here $X$ denotes dead end.

### 3.6.3 Hamiltonian Circuits Problem

- Hamiltonian circuit of a graph is a path that starts at a given vertex, visits each vertex in the graph exactly once and ends at the starting vertex.
- In Hamiltonian circuit problem, we have to find out the Hamiltonian circuits in the given graph.
- Backtracking approach can provide solution to Hamiltonian circuits problem.
- State space tree:
  (i) Put the starting vertex at level 0 in the tree.
  (ii) At level 1, create a child node for the root node for each remaining vertex that is adjacent to the first vertex.
  (iii) At each node in level 2, create a child node for each of the adjacent vertices that are not in the path from the root to this vertex, and so on.
  (iv) If at some level we encounter a dead end, we back track to the previous level, then move to the next choice and keep trying.

**Example**

Find Hamiltonian circuits in the following graph:

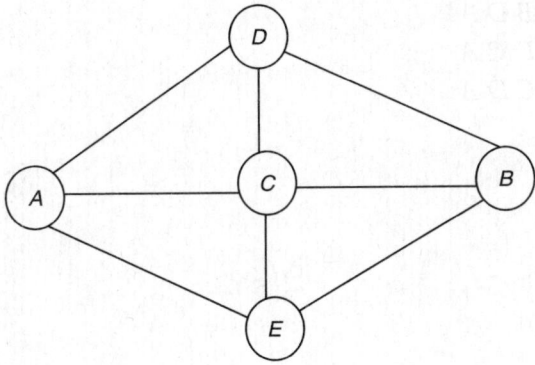

**Solution**

Following state space tree is used to find the Hamiltonian circuits in the given graph.

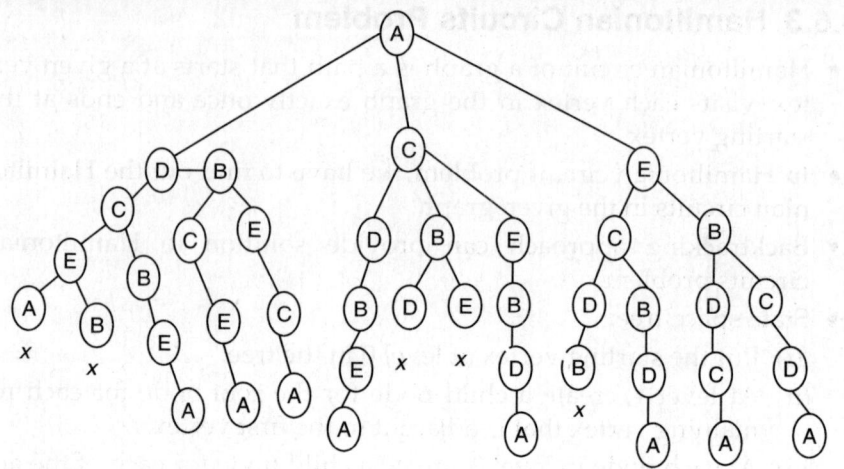

- Here X denotes dead end.

## Hamiltonian circuits:

1. $A\,D\,C\,B\,E\,A$
2. $A\,D\,B\,C\,E\,A$
3. $A\,D\,B\,E\,C\,A$
4. $A\,C\,D\,B\,E\,A$
5. $A\,C\,E\,B\,D\,A$
6. $A\,E\,C\,B\,D\,A$
7. $A\,E\,B\,D\,C\,A$
8. $A\,E\,B\,C\,D\,A$

# Chapter 4

# Miscellaneous Topics

## 4.1 COUNTING SORT

**Algorithm**

Counting-sort (A, B, k)

| | |
|---|---|
| Line 1 | let  C [0… k] be a new array |
| Line 2 | for  i ← 0 to k |
| Line 3 | C [i] ← 0 |
| Line 4 | for  j ← 1 to length [A] |
| Line 5 | C [A [j]] ← C [A [j]] + 1 |
| Line 6 | for  i ← 1 to k |
| Line 7 | C [i] ← C [i] + C [i – 1] |
| Line 8 | for  j ← length [A] down to 1 |
| Line 9 | B [C [A [j]]] ← A [j] |
| Line 10 | C [A [j]] ← C [A [j]] – 1 |

**Explanation**

- The inputs to the Counting-sort algorithm are an array $A [1 … n]$, an array $B [1…n]$ to hold the sorted output and '$k$', the largest element in array $A$.
- Line 1 indicates that $C [0 …k]$ is an auxiliary array that provides temporary working storage.
- Line 2 indicates the beginning of for loop that ends with Line 3. This loop is applicable for $i$ equals to 0 to $k$.
- Line 3 indicates that the for loop of Line 2 initially sets all the array entries to 0.

- Line 4 indicates the beginning of for loop that ends with Line 5. This loop is applicable for $j$ equals to 1 to $n$.
- Line 5 indicates that the value of $C[i]$ is incremented by 1 when the value of an input element $A[j]$ is $i$.

  After the execution of Line 5 array $C[i]$ contains the number of elements equal to $i$.

- Line 6 indicates the beginning of for loop that ends with Line 7. This loop is applicable for $i$ equals to 1 to $k$.
- Line 7 indicates that the new value of $C[i]$ is equal to the sum of previous values of $C[i]$ and $C[i-1]$.

  After the execution of Line 7 the array $C[i]$ contains the number of elements less than or equal to $i$.

- Line 8 indicates the beginning of for loop that ends with Line 10. This loop is applicable for $j$ equals to length $[A]$ down to 1.
- Line 9 indicates that each element $A[j]$ is placed into its correct sorted position in the output array $B$ with the help of auxiliary array $C$.
- Line 10 decrements the value $C[A[j]]$ by one which causes the next input element with a value equal to $A[j]$ to go to the position immediately before $A[j]$ in the output array and thus, maintaining the stability.
- Note that a sorting technique is said to be stable if the numbers with the same value appear in the output array in the same order as they do in the input array.

## Analysis

- The for loop of Lines 2-3 takes time $\Theta(k)$.
- The for loop of Lines 4-5 takes time $\Theta(n)$.
- The for loop of Lines 6-7 takes time $\Theta(k)$.
- The for loop of Lines 8-10 takes time $\Theta(n)$.
- Thus the overall time is $\Theta(k+n)$
- Counting sort is usually used when $k = O(n)$, in which case the running time is $\Theta(n)$.

## Example

Illustrate the operation of Counting-sort $(A, B, k)$ on the given array $A = 5, 2, 6, 4, 5, 3, 2, 3$

## Solution

| | 1 | 2 | 3 | 4 | 5 | 6 | 7 | 8 |
|---|---|---|---|---|---|---|---|---|
| $A =$ | 5 | 2 | 6 | 4 | 5 | 3 | 2 | 3 |

| | 1 | 2 | 3 | 4 | 5 | 6 | 7 | 8 |
|---|---|---|---|---|---|---|---|---|
| $B =$ | | | | | | | | |

$$k = 6$$

| | 0 | 1 | 2 | 3 | 4 | 5 | 6 |
|---|---|---|---|---|---|---|---|
| $C =$ | | | | | | | |

$$i = 0 \text{ to } 6$$

| | 0 | 1 | 2 | 3 | 4 | 5 | 6 |
|---|---|---|---|---|---|---|---|
| $C =$ | 0 | 0 | 0 | 0 | 0 | 0 | 0 |

$$j = 1 \text{ to } 8$$

for $\quad j = 1$

| | 0 | 1 | 2 | 3 | 4 | 5 | 6 |
|---|---|---|---|---|---|---|---|
| $C =$ | 0 | 0 | 0 | 0 | 0 | 1 | 0 |

for $\quad j = 2$

| | 0 | 1 | 2 | 3 | 4 | 5 | 6 |
|---|---|---|---|---|---|---|---|
| $C =$ | 0 | 0 | 1 | 0 | 0 | 1 | 0 |

for $\quad j = 3$

| | 0 | 1 | 2 | 3 | 4 | 5 | 6 |
|---|---|---|---|---|---|---|---|
| $C =$ | 0 | 0 | 1 | 0 | 0 | 1 | 1 |

for $\quad j = 4$

| | 0 | 1 | 2 | 3 | 4 | 5 | 6 |
|---|---|---|---|---|---|---|---|
| $C =$ | 0 | 0 | 1 | 0 | 1 | 1 | 1 |

for $\quad j = 5$

| | 0 | 1 | 2 | 3 | 4 | 5 | 6 |
|---|---|---|---|---|---|---|---|
| $C =$ | 0 | 0 | 1 | 0 | 1 | 2 | 1 |

for $\quad j = 6$

| | 0 | 1 | 2 | 3 | 4 | 5 | 6 |
|---|---|---|---|---|---|---|---|
| $C =$ | 0 | 0 | 1 | 1 | 1 | 2 | 1 |

for         $j = 7$

| $C =$ | 0 | 0 | 2 | 1 | 1 | 2 | 1 |
|---|---|---|---|---|---|---|---|
| | 0 | 1 | 2 | 3 | 4 | 5 | 6 |

for         $j = 8$

| $C =$ | 0 | 0 | 2 | 2 | 1 | 2 | 1 |
|---|---|---|---|---|---|---|---|
| | 0 | 1 | 2 | 3 | 4 | 5 | 6 |

for         $i = 1$ to 6

              $i = 1$

| $C =$ | 0 | 0 | 2 | 2 | 1 | 2 | 1 |
|---|---|---|---|---|---|---|---|
| | 0 | 1 | 2 | 3 | 4 | 5 | 6 |

              $i = 2$

| $C =$ | 0 | 0 | 2 | 2 | 1 | 2 | 1 |
|---|---|---|---|---|---|---|---|
| | 0 | 1 | 2 | 3 | 4 | 5 | 6 |

              $i = 3$

| $C =$ | 0 | 0 | 2 | 4 | 1 | 2 | 1 |
|---|---|---|---|---|---|---|---|
| | 0 | 1 | 2 | 3 | 4 | 5 | 6 |

              $i = 4$

| $C =$ | 0 | 0 | 2 | 4 | 5 | 2 | 1 |
|---|---|---|---|---|---|---|---|
| | 0 | 1 | 2 | 3 | 4 | 5 | 6 |

              $i = 5$

| $C =$ | 0 | 0 | 2 | 4 | 5 | 7 | 1 |
|---|---|---|---|---|---|---|---|
| | 0 | 1 | 2 | 3 | 4 | 5 | 6 |

              $i = 6$

| $C =$ | 0 | 0 | 2 | 4 | 5 | 7 | 8 |
|---|---|---|---|---|---|---|---|
| | 0 | 1 | 2 | 3 | 4 | 5 | 6 |

for         $j = 8$ to 1

              $j = 8$

| $B =$ | | | | 3 | | | | |
|---|---|---|---|---|---|---|---|---|
| | 1 | 2 | 3 | 4 | 5 | 6 | 7 | 8 |

| $C =$ | 0 | 0 | 2 | 3 | 5 | 7 | 8 |
|---|---|---|---|---|---|---|---|
| | 0 | 1 | 2 | 3 | 4 | 5 | 6 |

$j = 7$

| B = | | 2 | | 3 | | | | |
|---|---|---|---|---|---|---|---|---|
| | 1 | 2 | 3 | 4 | 5 | 6 | 7 | 8 |

| C = | 0 | 0 | 1 | 3 | 5 | 7 | 8 |
|---|---|---|---|---|---|---|---|
| | 0 | 1 | 2 | 3 | 4 | 5 | 6 |

$j = 6$

| B = | | 2 | 3 | 3 | | | | |
|---|---|---|---|---|---|---|---|---|
| | 1 | 2 | 3 | 4 | 5 | 6 | 7 | 8 |

| C = | 0 | 0 | 1 | 2 | 5 | 7 | 8 |
|---|---|---|---|---|---|---|---|
| | 0 | 1 | 2 | 3 | 4 | 5 | 6 |

$j = 5$

| B = | | 2 | 3 | 3 | | | 5 | |
|---|---|---|---|---|---|---|---|---|
| | 1 | 2 | 3 | 4 | 5 | 6 | 7 | 8 |

| C = | 0 | 0 | 1 | 2 | 5 | 6 | 8 |
|---|---|---|---|---|---|---|---|
| | 0 | 1 | 2 | 3 | 4 | 5 | 6 |

$j = 4$

| B = | | 2 | 3 | 3 | 4 | | 5 | |
|---|---|---|---|---|---|---|---|---|
| | 1 | 2 | 3 | 4 | 5 | 6 | 7 | 8 |

| C = | 0 | 0 | 1 | 2 | 4 | 6 | 8 |
|---|---|---|---|---|---|---|---|
| | 0 | 1 | 2 | 3 | 4 | 5 | 6 |

$j = 3$

| B = | | 2 | 3 | 3 | 4 | | 5 | 6 |
|---|---|---|---|---|---|---|---|---|
| | 1 | 2 | 3 | 4 | 5 | 6 | 7 | 8 |

| C = | 0 | 0 | 1 | 2 | 4 | 6 | 7 |
|---|---|---|---|---|---|---|---|
| | 0 | 1 | 2 | 3 | 4 | 5 | 6 |

$j = 2$

| B = | 2 | 2 | 3 | 3 | 4 | | 5 | 6 |
|---|---|---|---|---|---|---|---|---|
| | 1 | 2 | 3 | 4 | 5 | 6 | 7 | 8 |

| C = | 0 | 0 | 0 | 2 | 4 | 6 | 7 |   |
|-----|---|---|---|---|---|---|---|---|
|     | 0 | 1 | 2 | 3 | 4 | 5 | 6 |   |

$$j = 1$$

| B = | 2 | 2 | 3 | 3 | 4 | 5 | 5 | 6 |
|-----|---|---|---|---|---|---|---|---|
|     | 1 | 2 | 3 | 4 | 5 | 6 | 7 | 8 |

| C = | 0 | 0 | 0 | 2 | 4 | 5 | 7 |   |
|-----|---|---|---|---|---|---|---|---|
|     | 0 | 1 | 2 | 3 | 4 | 5 | 6 |   |

## 4.2 RADIX SORT

### Algorithm

Radix-Sort (A, d)

Line 1        for  i ← 1 to d

Line 2              use a stable sort to sort array A on digit i

### Explanation

- Radix-sort $(A, d)$ procedure assumes that each element in the input array $A$ has $d$ digits.
- Line 1 indicates the beginning of for loop that ends with Line 2. This for loop is applicable for $i$ equals to 1 to $d$.
- Line 2 indicates that a stable sort like counting sort can be used to sort array $A$ on digit $i$.

### Analysis

- Consider $n$, $d$-digit numbers in which each digit can take on up to $k$ possible values.
- If the stable sort used in Radix-Sort procedure takes $\Theta (n + k)$ time then the Radix-sort procedure itself takes $\Theta (d (n + k))$ time.

### Example

Sort the following words using Radix-Sort procedure.

CAT, MAT, RAT, HEN, PEN, GUN, RUN, FUN, COW, DOG, FAT, BAT, DOT, COT, EAT, ANT, INK, JUG, NUT, POT.

### Solution

|     |     |     |     |
|-----|-----|-----|-----|
| CAT | DOG | CAT | ANT |
| MAT | JUG | MAT | BAT |

| RAT | INK | RAT | CAT |
|-----|-----|-----|-----|
| HEN | HEN | FAT | COT |
| PEN | PEN | BAT | COW |
| GUN | GUN | EAT | DOG |
| RUN | RUN | HEN | DOT |
| FUN | FUN | PEN | EAT |
| COW | CAT | INK | FAT |
| DOG | MAT | ANT | FUN |
| FAT | RAT | DOG | GUN |
| BAT | FAT | DOT | HEN |
| DOT | BAT | COT | INK |
| COT | DOT | POT | JUG |
| EAT | COT | COW | MAT |
| ANT | EAT | JUG | NUT |
| INK | ANT | GUN | PEN |
| JUG | NUT | RUN | POT |
| NUT | POT | FUN | RAT |
| POT | COW | NUT | RUN |

- In the above example the for loop runs from 1 to 3.
  For $i$ equals to 1, the left most column is the input.
  For $i$ equals to 2, the middle column is the input.
  In the end for $i$ equals to 3, the third column is the input.
- The highlighted list in each column indicates the digit position sorted on, to produce each list from the previous one.
- After the execution of for loop we obtain a sorted list of words (right most column).

## 4.3 BUCKET SORT

### Algorithm

```
Bucket-Sort (A)
Line 1      let  B [0... n – 1] be a new array
Line 2      n ← length [A]
Line 3      for  i ← 0 to n – 1
Line 4              make B [i] an empty list
Line 5      for  i ← 1 to n
Line 6              insert A [i] into list B [ ⌊nA[i]⌋ ]
Line 7      for  i ← 0 to n – 1
```

Line 8                    sort list B [i] with insertion sort
Line 9          concatenate the lists B [0], B [1]... B [n – 1]
                   together in order

## Explanation

- The input to the Bucket-Sort ($A$) algorithm is an array $A$ having $n$ elements and its each element A[$i$] satisfies the relation $0 \leq A[i] < 1$.
- Line 1 indicates that B [0... $n – 1$] is an auxiliary array of linked lists.
- Line 2 indicates that $n$ equals to the length of array $A$.
- Line 3 indicates the beginning of for loop that ends with Line 4. This for loop is applicable for $i$ equals to 0 to $n – 1$.
- Line 4 makes B [$i$] an empty list.
- Line 5 indicates the beginning of for loop that ends with Line 6. This loop is applicable for $i$ equals to 1 to $n$.
- Line 6 inserts A[$i$] into list B [$\lfloor nA[i] \rfloor$].
- Line 7 indicates the beginning of for loop that ends with Line 8. The for loop is applicable for $i$ equals to 0 to $n – 1$.
- Line 8 indicates that list B[$i$] is sorted using insertion sort technique.
- Line 9 concatenates the lists B[0], B[1]... B[$n – 1$] together in order.

## Analysis

- The worst case running time for bucket sort is $\Theta (n^2)$.

## Example

Apply the operation of Bucket-Sort ($A$) on the following array:

$A = 0.42, 0.69, 0.78, 0.53, 0.31, 0.36, 0.11, 0.24, 0.62, 0.39$

## Solution

| | A |
|---|---|
| 1 | 0.42 |
| 2 | 0.69 |
| 3 | 0.78 |
| 4 | 0.53 |
| 5 | 0.31 |
| 6 | 0.36 |
| 7 | 0.11 |
| 8 | 0.24 |
| 9 | 0.62 |
| 10 | 0.39 |

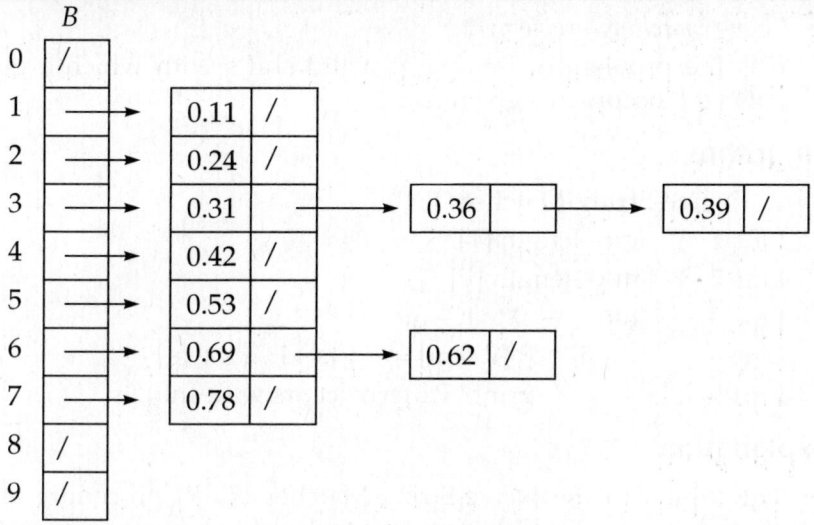

Apply insertion sort to sort the elements

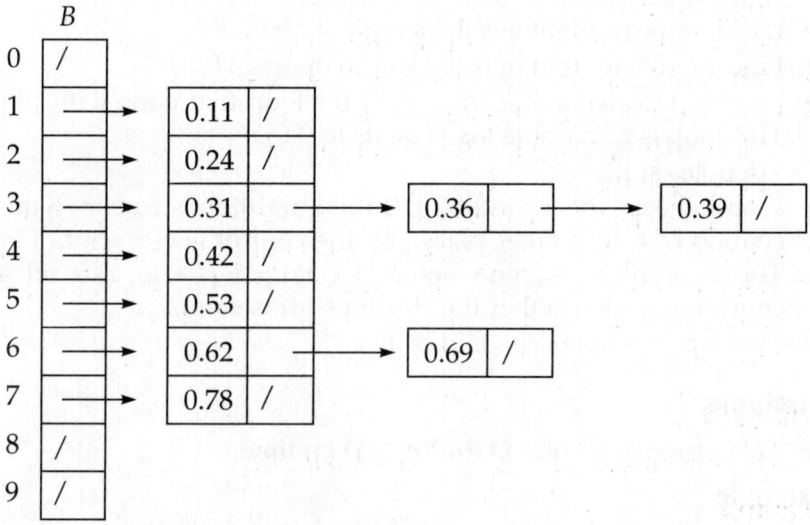

## 4.4 STRING MATCHING

- Assume that the text is an array $T[1...n]$ of length $n$ and that the pattern is an array $P[1... m]$ of length $m \leq n$.
- Also assume that the elements of the character array $P$ and $T$ are drawn from the finite alphabet $\Sigma$.
- The pattern $P$ occurs with shift $s$ in text $T$, if $0 \leq s \leq n - m$ and $T[s+1 ... s+m] = P[1 ...m]$.
- If $P$ occurs with shift $s$ in $T$, then it is called a valid shift else $s$ is called an invalid shift.

- *String matching problem*:
  It is the problem of finding all valid shifts with which a given pattern $P$ occurs in a given text $T$.

## Algorithm

> Naïve-String-Matcher (T, P)

Line 1     n ← length [T]

Line 2     m ← length [P]

Line 3     for s ← 0 to n − m

Line 4        if P [1... m] = = T [s +1 ... s + m]

Line 5        print "Pattern occurs with shift" s

## Explanation

- The inputs to the Naïve-String-Matcher $(T, P)$ procedure are a character array $T$ which stores the text and a character array $P$ that stores the pattern.
- Line 1 indicates that n is the length of array $T$.
- Line 2 indicates that m is the length of array $P$.
- Line 3 indicates the beginning of for loop that ends with Line 5. This loop is applicable for $s$ equals to 0 to $n - m$. $s$ denotes shift.
- Line 4 checks the *if* condition. If this condition is true then the execution of Line 5 takes place else the control goes back to Line 3. The *if* condition is true when $P [1... m] = T [s + 1 ....s + m]$. This condition tests whether the current shift is valid.
- Line 5 prints out each valid shift $s$.

## Analysis

- This procedure takes $O ((n - m + 1) m)$ time.

## Example

Show the comparisons made by Naïve-String-Matcher for the pattern $P = 001$ in the text $T = 000010100100101$:

## Solution

$$n = 15$$
$$m = 3$$

for $s = 0$

$$P [1] = = T [0 + 1]$$
$$P [2] = = T [0 + 2]$$
$$P [3] \neq T [0 + 3]$$

So, *if* condition fails

for $s = 1$

$$P[1] == T[1+1]$$
$$P[2] == T[1+2]$$
$$P[3] \neq T[1+3]$$

*if* condition fails

for $s = 2$

$$P[1] == T[2+1]$$
$$P[2] == T[2+2]$$
$$P[3] == T[2+3]$$

Pattern occurs with shift 2

for $s = 3$

$$P[1] == T[3+1]$$
$$P[2] \neq T[3+2]$$

for $s = 4$

$$P[1] \neq T[4+1]$$

for $s = 5$

$$P[1] == T[5+1]$$
$$P[2] \neq T[5+2]$$

for $s = 6$

$$P[1] \neq T[6+1]$$

for $s = 7$

$$P[1] == T[7+1]$$
$$P[2] == T[7+2]$$
$$P[3] == T[7+3]$$

Pattern occurs with shift 7

for $s = 8$

$$P[1] == T[8+1]$$
$$P[2] \neq T[8+2]$$

for $s = 9$

$$P[1] \neq T[9+1]$$

for $s = 10$

$$P[1] == T[10+1]$$
$$P[2] == T[10+2]$$
$$P[3] == T[10+3]$$

Pattern occurs with shift 10

for $s = 11$

$$P[1] == T[11+1]$$
$$P[2] \neq T[11+2]$$

for $s = 12$

$$P[1] \neq T[12+1]$$

Now for loop terminates.

## 4.5 METHODS TO SOLVE RECURRENCE RELATIONS

### 4.5.1 The Recursion Tree Method

- In a recursion tree, each node represents the cost of a single sub problem somewhere in the set of recursive function invocations.
- The costs within each level of the tree are summed up to obtain a set of pre-level costs and then all the pre-level costs are summed up to determine the total cost of all levels of recursion.

**Example**

Construct a recursion tree for the recurrence $T(n) = 2T(n/2) + cn^2$

**Solution**

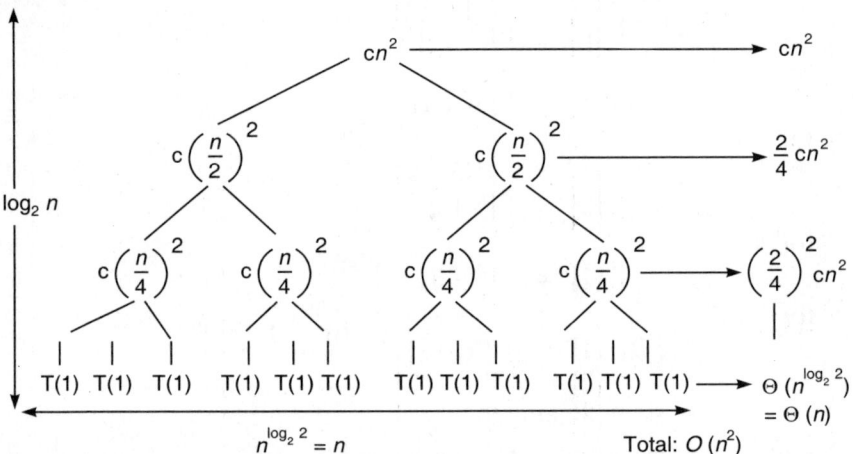

- Here the subproblem sizes decrease by a factor of 2.
- The subproblem size for a node at depth $i$ is $\dfrac{n}{2^i}$, therefore, the sub problem size hits $n = 1$, when $\dfrac{n}{2^i} = 1$ or when $i = \log_2 n$. Thus the tree has $\log_2 n + 1$ levels.
- Each level has two times more nodes than the level above, so the number of nodes at depth $i$ is $2^i$.
- The costs over all levels are added-up to determine the cost of the entire tree:

$$T(n) = cn^2 + \left(\frac{2}{4}\right) cn^2 + \left(\frac{2}{4}\right)^2 cn^2 + \ldots + \left(\frac{2}{4}\right)^{\log_2 n - 1} cn^2$$

$$+ \Theta\left(n^{\log_2 2}\right)$$

$$= \sum_{i=0}^{\log_2 n - 1} \left(\frac{2}{4}\right)^i cn^2 + \Theta(n)$$

$$< \sum_{-}^{\infty} \left( - \right) cn^2 + \Theta(n)$$

$$= \frac{1}{1 - \left( \dfrac{2}{4} \right)} cn^2 + \Theta(n)$$

$$= 2\, cn^2 + \Theta(n)$$

$$= O(n^2)$$

**Example**

Construct a recursion tree for the recurrence

$$T(n) = T(2n/3) + T(n/3) + cn$$

**Solution**

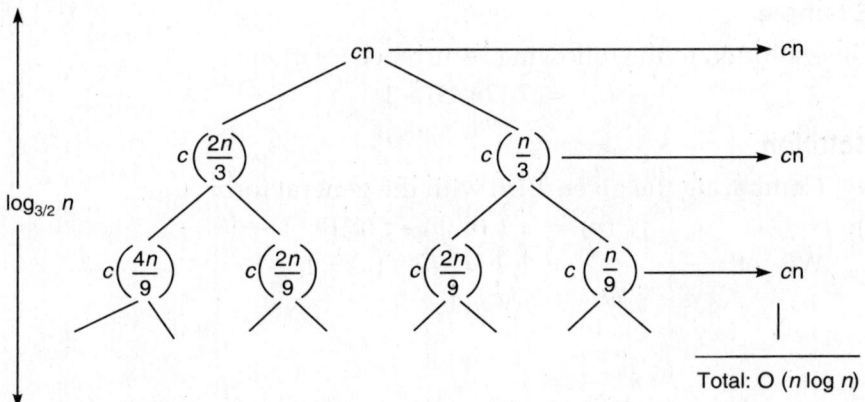

- Here the recursion tree is not the complete binary tree.

  Therefore,   total cost $T(n) = (\log_{3/2} n)(cn)$

  $$= O(n \lg n)$$

## 4.5.2 The Master Method

- Let $a \geq 1$ and $b > 1$ be constants, $f(n)$ be a function, and $T(n)$ be defined on the non-negative integers by the recurrence

  $$T(n) = a\, T(n/b) + f(n)$$

  where, $\dfrac{n}{b}$ can be either $\left\lfloor \dfrac{n}{b} \right\rfloor$ or $\left\lceil \dfrac{n}{b} \right\rceil$.

- To get the solution of the recurrence $T(n)$, the function $f(n)$ is compared with the function $n^{\log_b a}$. After the comparison one of the following three cases will be applied to find the solution.

Case (i): If function $n^{\log_b a}$ is larger, then the solution is
$$T(n) = \Theta(n^{\log_b a}).$$
In otherwords,
If $f(n) = O(n^{\log_b a - \epsilon})$ for some constant $\epsilon > 0$, then
$$T(n) = \Theta(n^{\log_b a}).$$

Case (ii): If functions $n^{\log_b a}$ and $f(n)$ are of same size, then
$$T(n) = \Theta(n^{\log_b a} \lg n) = \Theta(f(n) \lg n).$$

Case (iii): If function $f(n)$ is larger, then the solution is $T(n) = \Theta(f(n))$
In other words,
If $f(n) = \Omega(n^{\log_b a + \epsilon})$ for some constant $\epsilon > 0$, and
if a $f(n/b) \le c f(n)$ for some constant $c < 1$ and all sufficiently large $n$, then $T(n) = \Theta(f(n))$.

## Example

Give solution to the following recurrence relation
$$T(n) = T(2n/3) + 1$$

## Solution

Comparing the given T $(n)$ with the general form
$$[T(n) = a T(n/b) + f(n)]$$
We get,      $a = 1, b = 3/2, f(n) = 1$
$$n^{\log_b a} = n^{\log_{3/2} 1}$$
$$= n^0$$
$$n^{\log_b a} = 1$$
Since $f(n)$ and $n^{\log_b a}$ are equal, case (ii) will be applied.
Therefore,      $T(n) = \Theta(\lg n)$

## Example

Give solution to the given recurrence relation
$$T(n) = 9 T(n/3) + n$$

## Solution

Comparing the given relation with the general form
$$[T(n) = a T(n/b) + f(n)]$$
We get,            $a = 9, b = 3, f(n) = n$
$$n^{\log_b a} = n^{\log_3 9}$$
$$n^{\log_b a} = n^2$$
Since, $n^{\log_b a}$ is larger than $f(n)$, case 1 will be applied.
Therefore,      $T(n) = \Theta(n^2)$.

# Bibliography

1. Cormen TH, Leiserson CE, Rivest RL, Stein C. *Introduction to Algorithms*, 3rd edn, MIT Press: Cambridge, MA, 2009.
2. Kleinberg J, Tardoz E. *Algorithm Design*. Addison-Wesley:MA, USA 2006.
3. Horowitz E, Sahni S, Rajasekaran S. *Fundamentals of Computer Algorithms*. 2nd edn, Universities Press: India 2009.
4. Langsam Y, Angenstein MJ, Tenenbaum AM. *Data Structures using C and C++*. 2nd edn, Pearson: NJ, USA, 2010.
5. Lipschutz S. *Data Structures*. McGraw-Hill: NY, 2010.
6. Levitin A. *Introduction to the Design and Analysis of Algorithms*. Addison-Wesley: MA, USA, 2012.

# INDEX

255